Six-Minute Solutions

for Chemical PE Exam Problems

Marta Vasquez, PhD, PE
Robert R. Zinn

Professional Publications, Inc. • Belmont, California

> **Benefit by Registering This Book with PPI**
>
> - Get book updates and corrections
> - Hear the latest exam news
> - Obtain exclusive exam tips and strategies
> - Receive special discounts
>
> Register your book at **www.ppi2pass.com/register**.
>
> **Report Errors and View Corrections for This Book**
>
> PPI is grateful to every reader who notifies us of a possible error. Your feedback allows us to improve the quality and accuracy of our products. You can report errata and view corrections at **www.ppi2pass.com/errata**.

SIX-MINUTE SOLUTIONS FOR CHEMICAL PE EXAM PROBLEMS

Current printing of this edition: 4

Printing History

edition number	printing number	update
1	2	Minor corrections.
1	3	Minor corrections.
1	4	Minor corrections. Copyright update.

Copyright © 2011 by Professional Publications, Inc. (PPI). All rights reserved. No part of this publication may be reproduced, stored in a retrieval system, or transmitted, in any form or by any means, electronic, mechanical, photocopying, recording, or otherwise, without the prior written permission of the publisher.

Printed in the United States of America.

PPI
1250 Fifth Avenue, Belmont, CA 94002
(650) 593-9119
www.ppi2pass.com

Library of Congress Cataloging-in-Publication Data
Vasquez, Marta
    Six-minute solutions for chemical PE exam problems / Marta Vasquez, Robert R. Zinn.
       p. cm.
    Includes bibliographical references.
    ISBN: 978-1-59126-012-7
    1. Chemical engineering--Problems, exercises, etc. I. Title: 6-minute solutions for chemical PE exam problems. II. Zinn, Robert R. III. Title.
TP168.V37 2004
660′.076--dc22
                                            2004040128

Table of Contents

ABOUT THE AUTHORS iv

PREFACE AND ACKNOWLEDGMENTS v

INTRODUCTION
    Exam Format vii
    This Book's Organization vii
    How to Use This Book vii

REFERENCES ix

NOMENCLATURE xi

PROBLEMS
    Mass/Energy Balances and Thermodynamics 1
    Fluids 8
    Heat Transfer 13
    Mass Transfer 16
    Kinetics 19
    Plant Design and Operation 21

SOLUTIONS
    Mass/Energy Balances and Thermodynamics 27
    Fluids 56
    Heat Transfer 71
    Mass Transfer 82
    Kinetics 98
    Plant Design and Operation 108

About the Authors

Marta Vasquez, PhD, PE, is an environmental chemist with the Louisiana Department of Environmental Quality (LDEQ). She holds a BS degree in chemical engineering from Antioquia University in Medellin, Colombia, an MS degree in organic chemistry from Del Valle University, Cali, Colombia, and a doctorate in analytical chemistry from Louisiana State University. In 1971, she joined the faculty of Caldas University in Colombia and attained the rank of Associate Professor of organic chemistry prior leaving in 1984 to pursue her doctoral degree from Louisiana State University. Since earning her doctorate in August 1989, Dr. Vasquez has worked in the field of environmental chemistry. Currently, with LDEQ, she works with the portions of the state and federal regulations that are applicable to petrochemical industries such as refineries, polymer manufacturers, and paper mills.

Dr. Vasquez is a co-author of various scientific papers published in *Phytochemistry* in 1988, 1989, 1990, and 1992 and in *Acta Crystallographica* in 1990 and 1992. She is also a co-author, with R.R. Zinn and others, of two unpublished reviews on sesquiterpene lactones. In addition she collaborated on the writing of a scientific paper published in the journal *Magnetic Resonance in Chemistry* in 1990. She is a registered professional engineer in Louisiana.

Robert R. Zinn received his BS degree in chemistry, with a minor in mathematics, from Midwestern University in Wichita Falls, Texas, in 1970. He has worked in the field of electronic component manufacturing, electronic service, computer repair, instrument repair, and computer programming since 1969. He currently provides technical support, as systems manager, for faculty and staff at the Louisiana State University Chemistry Department, where he has served since 1981.

He is the author of "The Turtle that Learned to Fly" and "A Song for Heroes, Unsung," published in *Our Journey* in May 1995. He is a co-author of the papers "Optogalvanic Transients in a Neon RF Discharge" and "Optogalvanic Effects as a Probe of Plasma Processes," published in the *Journal of Physical Chemistry* in 1994 and 1995. He also is a co-author, with M. Vasquez and others, of two unpublished reviews on sesquiterpene lactones.

Preface and Acknowledgments

The Principles and Practice of Engineering examination (PE exam) for chemical engineering, prepared by the National Council of Examiners for Engineering and Surveying (NCEES), is developed from sample problems submitted by educators and professional engineers representing consulting, government, and industry. PE exams are designed to test examinees' understanding of both conceptual and practical engineering concepts. Problems from past exams are not available from NCEES or any other source. However, NCEES does identify the general subject areas covered on the exam.

The topics covered in *Six-Minute Solutions for Chemical PE Exam Problems* coincide with those subject areas identified by NCEES for the chemical PE exam. Included among these problem topics are mass/energy balances and thermodynamics, fluids, heat transfer, mass transfer, kinetics, and plant design and operation.

The problems presented in this book are representative of the type and difficulty of problems you will encounter on the PE exam. The book's problems are both conceptual and practical, and they are written to provide varying levels of difficulty. Though you probably won't encounter problems on the exam exactly like those presented here, reviewing these problems and solutions will increase your familiarity with the exam problems' form, content, and solution methods. This preparation will help you considerably during the exam.

Problems and solutions have been carefully prepared and reviewed to ensure that they are appropriate and understandable, and that they were solved correctly. If you find errors or discover an alternative, more efficient way to solve a problem, please bring it to PPI's attention so your suggestions can be incorporated into future editions. You can report errors and keep up with the changes made to this book, as well changes to the exam, by logging on to PPI's website at **www.ppi2pass.com** and clicking on "Errata."

The completion of a project, such as this book, requires the efforts of many people. We owe many debts of gratitude in this regard. We wish to thank, first and foremost, the members of our close and extended families who gave their love and support during the time it took to write this book.

We wish to thank the PPI production and editorial staff. Several individuals, due to their keen critical insights, have been especially important in the development of this book. In particular we want to thank Sarah Hubbard, Aline Magee, Kathleen Sullivan, and Heather Kinser.

In addition, thanks go to Dr. Emmanuel Wada, who was a source of technical information. We offer our sincere appreciation to Larry Rotter because his knowledge and expertise were key to improving the solutions to the problems presented here. Thanks go also to Dr. John Richards and Chuck Simchick. They made significant contributions to this book.

Although a mere acknowledgment is insufficient, we wish to recognize the support of the editor-in-chief of PPI, Michael Lindeburg. He inspired us, as we admire his great contributions and accomplishments in this field.

We certainly owe thanks to those who challenged us, corrected us, and motivated us. Finally, we want to thank all those who found errors and ambiguity in the text and presented their opinions and suggestions, especially those behind the scenes, who spent much effort and time.

Marta Vasquez, PhD, PE
Robert R. Zinn

Introduction

EXAM FORMAT

The Principles and Practice of Engineering examination (PE exam) in chemical engineering is an 8-hour exam divided into a morning and an afternoon session.

The morning and afternoon sessions each include 40 problems from the six chemical engineering subdisciplines (mass/energy balances and thermodynamics, fluids, heat transfer, mass transfer, kinetics, and plant design and operation). All problems are multiple choice. They include a problem statement with all required defining information, followed by four logical choices. Only one of the four options is correct. Nearly every problem is completely independent of all others, so an incorrect choice on one problem typically will not carry over to subsequent problems.

Topics and the approximate distribution of problems on the chemical exam are as follows.

- Mass/Energy Balances and Thermodynamics: approximately 24% of exam problems
- Fluids: approximately 17% of exam problems
- Heat Transfer: approximately 16% of exam problems
- Mass Transfer: approximately 13% of exam problems
- Kinetics: approximately 11% of exam problems
- Plant Design and Operation: approximately 19% of exam problems

Passing the test requires a score of at least 70%. This implies that you must correctly answer 56 of the 80 problems, unless the examination committee decides that a question is defective and discards it. In other words, if you take more than approximately $8^{1}/_{2}$ minutes to solve each problem, it is possible that you will NOT pass, *even if all of your answers are correct*. This makes practice at working problems vital to passing the test.

For further information and tips on how to prepare for the chemical environmental engineering PE exam, consult the *Chemical Engineering Reference Manual* or PPI's website, **www.ppi2pass.com**.

THIS BOOK'S ORGANIZATION

Six-Minute Solutions for Chemical PE Exam Problems is organized into six sections, one for each subject on the exam.

Most of the problems are quantitative, requiring calculations to arrive at a correct solution. A few are non-quantitative. Some problems will require a little more than 6 minutes to answer and others a little less. In order to answer all the questions, you should aim to complete 80 problems in 480 minutes (8 hours), or on average, spend 6 minutes per problem.

HOW TO USE THIS BOOK

In *Six-Minute Solutions for Chemical PE Exam Problems*, each problem statement, with its supporting information and answer choices, is presented in the same format as the problems encountered on the PE exam. The solutions are presented in a step-by-step sequence to help you follow the logical development of the correct solution and to provide examples of how you may want to approach your solutions as you take the PE exam.

Each problem includes a hint to provide direction in solving the problem. In addition to the correct solution, you will find an explanation of the faulty solutions leading to the three incorrect answer choices. The incorrect solutions are intended to represent common mistakes made when solving each type of problem. These may be simple mathematical errors, such as failing to square a term in an equation, or more serious errors, such as using the wrong equation.

To optimize your study time and obtain the maximum benefit from the practice problems, consider the following suggestions.

1. Complete an overall review of the problems and identify the subjects that you are least familiar with. Work a few of these problems to assess your general understanding of the subjects and to identify your strengths and weaknesses.

2. Locate and organize relevant resource materials. (See the References section of this book for guidance.) As you work problems, some of these resources will emerge as more useful to you than others. You will want to have these on hand when taking the PE exam.

3. Work the problems in one subject area at a time, starting with the subject areas that you have the most difficulty with.

4. When possible, work problems without utilizing the hint. Always attempt your own solution before looking at the solutions provided in the book. Use the solutions to check your work or to provide guidance in finding solutions to the more difficult problems. Use the incorrect solutions to help identify pitfalls and to develop strategies to avoid them.

5. Use each subject area's solutions as a guide to understanding general problem-solving approaches. Although problems identical to those presented in *Six-Minute Solutions for Chemical PE Exam Problems* will not be encountered on the PE exam, the approach to solving problems will be the same.

Solutions presented for each example problem may represent only one of several methods for obtaining a correct answer. Although most of these problems have unique solutions, alternative problem-solving methods may produce a different, but nonetheless appropriate, answer.

Reference books often format their equations with "unitless factors." It is "understood" that the particular form of the equation is ONLY valid if the rest of the variables in the equation are expressed in the proper units. In this book you will find that the implied units for these unitless factors are shown explicitly in order to emphasize the importance of proper unit cancellation.

References

The minimum recommended library for the chemical exam consists of PPI's *Chemical Engineering Reference Manual*. You may also find the following references helpful in completing some of the problems in *Six-Minute Solutions for Chemical PE Exam Problems*.

Blackwell, Wayne W. *Chemical Process Design on a Programmable Calculator*. McGraw-Hill.

Cochran, Thomas W. "Simplifying Piping Network Analysis." *Chemical Engineering* (April 1995): 104–106.

Crane Co. "Flow of Fluids Through Valves, Fittings, and Pipe" (Technical Paper No. 410).

Dean, John A. *Lange's Handbook of Chemistry*. McGraw-Hill.

Levenspiel, Octave. *Chemical Reaction Engineering*. John Wiley and Sons.

Lindeburg, Michael R. *Chemical Engineering Reference Manual for the PE Exam*. Professional Publications.

McCabe, Warren L., Julian C. Smith, and Peter Harriott. *Unit Operations of Chemical Engineering*. McGraw-Hill.

Metcalf & Eddy, Inc. *Wastewater Engineering: Treatment, Disposal, and Reuse*. McGraw-Hill.

Ostler, Neal K., ed. *Prentice Hall's Environmental Technology Series, Volume I: Introduction to Environmental Technology*. Prentice Hall.

Peavy, Howard S., et al. *Environmental Engineering*. McGraw-Hill.

Perry, Robert H., and Don W. Green. *Perry's Chemical Engineers' Handbook*. McGraw-Hill.

Sawyer, Clair N., et al. *Chemistry for Environmental Engineering*. McGraw-Hill.

Smith, J.M., et al. *Introduction to Chemical Engineering Thermodynamics*. McGraw-Hill.

Nomenclature

A	area	ft²	m²
A	hypothetical component mass flow rate	tons/day	kg/d
A	mass of component A in the feed	lbm	kg
AF	absorption factor	–	–
b	specific biomass rate	day⁻¹	d⁻¹
B	bottoms flow rate	lbm/hr	kg/h
B	hypothetical component mass flow rate	tons/day	kg/d
B	mass of component B in the feed	lbm	kg
$\bar{c}$	average specific heat capacity	Btu/lbm-°F	kJ/kg·°C
c	refrigeration capacity	tons	–
c	specific heat capacity	Btu/lbm-°F	kJ/kg·°C
$\bar{C}$	average molar heat capacity	Btu/lbmol-°F	kJ/kmol·°C
C	cost	$	$
C	concentration	lbm/ft³, lbmol/ft³	mg/L, mol/L, kg/m³
C	discharge coefficient	–	–
C	flow coefficient	–	–
C	hypothetical component mass flow rate	tons/day	kg/d
C	orifice coefficient	–	–
C	rate constant	lbmol-in²/lbm-lbf-hr	kmol/kg·atm·h
CE	energy actually recovered from flue gas	Btu	kJ
d	diameter	in	m
D	amount of distillate	lbm, lbmol	kg, mol
D	diameter	ft	m
D	distillate flow rate	lbm/hr, lbmol/min	kg/h, mol/min
DRE	destruction and removal efficiency	–	–
E	amount of extract	lbm, ton	kg
E	energy change rate	Btu/hr	kJ/h
E	extraction factor	–	–
E	flow rate of extract	tons/day	kg/d
E	fractional energy recovery	–	–
E	molar activation energy	Btu/lbmol	kJ/mol
E	splitter ratio	–	–
EW	equivalent weight	lbm/lbmol	kg/mol
f	fraction of flooding velocity	–	–
f	friction factor	–	–
F	amount of feed	lbmol	kmol
F	component flow rate in feed	–	–
F	heat exchanger correction factor	–	–
F	mass flow rate of feed	tons/day	kg/d
F	molar flow rate	lbmol/hr	mol/h
F	reactor inlet	lbmol/hr	kmol/h
g	acceleration due to gravity	ft/sec²	m/s²
G	gas flow rate	ft³/min	m³/min
G	molar gas flow rate	lbm/hr	g/min
G	molar Gibbs free energy	Btu/mol	kJ/mol
G	solute-free molar gas flow rate	lbmol/hr	kgmol/h

g_c	gravitational constant	lbm-ft/lbf-sec^2	kg·m/N·s^2
h	convection coefficient	Btu/ft^2-hr-°R	W/m^2·K
h	enthalpy	Btu	cal, kJ
h	heat-transfer coefficient	Btu/ft^2-hr-°R	W/m^2·°C
h	height or length of vessel	ft	m
h	layer thickness	ft	m
h	organic phase	–	–
h	pressure head	ft	m
h	pump head	ft	m
h	radiation coefficient	Btu/ft^2-hr-°R	W/m^2·K
h	specific enthalpy	Btu/lbm	kJ/kg
h	static head	ft	m
h	velocity	ft	m
H	molar enthalpy	Btu/mol	kJ/mol
H	volumetric heat of combustion	Btu/ft^3	kJ/m^3
$\dot{H}$	rate of enthalpy change	Btu/min	kJ/s
HC	hydrocarbon layer	–	–
HD	heat duty	hp	kW
k	air isentropic exponent	–	–
k	chemical reaction equilibrium constant	–	–
k	integration constant	–	–
k	reaction-rate constant	lbmol/ft^3/sec, hr^{-1}, ft^3/lbmol-sec	mol/cm^3/s, h^{-1}, L/mol·s
k	thermal conductivity	Btu/hr-ft-°F	W/m·°C
k	vapor-liquid equilibrium constant	–	–
K	concentration equilibrium constant	lbmol/ft^3, ft^5/lbmol^2	mol/L, L^2/mol^2
K	equilibrium partition coefficient	–	–
K	equilibrium ratio for vapor-liquid equilibria	–	–
K	flow resistance coefficient	sec^2/ft^5	s^2/m^5
K	ion product of water	–	–
K	overall gas mass-transfer coefficient	lbmol/ft^2-hr	mol/s·cm^2
K	overall liquid mass-transfer coefficient	lbmol/ft^2-hr	mol/s·cm^2
K	resistance coefficient	sec^2/ft^5	s^2/m^5
L	equivalent length	ft	m
L	height	ft	m
L	length	ft	m
L	liquid flow rate	ft^3/min	m^3/min
L	liquid level	ft	m
L	liquid molar flow rate	lbm/hr	kmol/h
L	solute-free liquid molar flow rate	lbmol/hr	mol/h
L	superficial liquid velocity	ft/sec	cm/s
m	mass	lbm	kg
m	slope of equilibrium curve	min^{-1}	min^{-1}
$\dot{m}$	mass flow rate	lbm/min	kg/min
M	mass of mixture in solid-liquid extraction	lbm	kg
M	ratio of initial concentrations	–	–
MW	molecular weight	lbm/lbmol	kg/mol
n	number	–	–
N	number of plates	–	–
NPSHA	net positive suction head available	ft	m
p	pressure	atm, lbf/in^2, in H$_2$O	mm Hg, kPa
P	power	ft-lbf/sec, hp	W, J/s
P	product production	lbm, lbmol	kg, kmol
q	caloric energy	Btu	kJ

q	specific heat transfer	Btu/lbm	kJ/kg
Q	compressor ratio	–	
Q	heat	Btu	kJ/kg, cal, J
Q	heat-transfer rate	Btu/hr	kJ/s, W
Q	liquid volumetric flow rate	ft^3/sec, gal/min	L/min, m^3/d
Q	reaction rate	lbmol/ft^3-sec	mol/cm^3·s, mol/L·min
r	reaction rate	lbmol/hr	mol/min
R	amount of raffinate	lbm	kg
R	flow rate of raffinate	lbm/day	kg/d
R	mass of raffinate phase	lbm	kg
R	range	–	
R	thermal resistance to heat transfer	hr-°R/Btu	K/W
R	universal gas constant	lbf-ft^3/lbmol-in^2-°R	mm Hg·cm^3/mol·K, J/mol·K
Re	Reynolds number	–	
s	entropy	Btu/lbm	kJ/kg
s	specific entropy	Btu/lbm-°R	kJ/kg·K
S	molar entropy	Btu/lbmol-°F	J/mol·K
S	solvent mass flow rate	tons/day	kg/d
SG	specific gravity	–	–
t	time	min	min
T	temperature	°F, °R	K, °C
TAS	net total annual savings	\$/yr	\$/yr
TCI	total capital investment	\$	\$
U	flooding velocity	ft^3/sec	L/s
U	overall heat-transfer coefficient	Btu/ft^2-hr-°F	W/m^2·°C
U	underflow flow rate	lbm/hr	mol/h
u_s	superficial gas velocity in a packed column	ft/sec	m/s
v	flow rate	ft/sec	m/s
V	overflow flow rate	ft^3/hr	m^3/h, L/h
V	vapor flow rate	ft^3/hr	m^3/h, L/h
V	volume	ft^3	cm^3, L
$\dot{V}$	gaseous volumetric flow rate	ft^3/min	m^3/min, L/min
w	mass fraction	–	–
W	bottom flow rate	lbm/hr	kg/h
W	moles of liquid in a batch still	lbmol	mol
W	rate of work	Btu	kW/h
W	mass flow rate of water	tons/day	kg/d
W	width of film	ft	cm
W	work	Btu/hr	kW
x	liquid concentration in equilibrium with gas at its bulk partial pressure	lbm/ft^3	mol/L
x	mass or mole fraction	–	–
X	abscissa in the Eckert's flooding curve	–	–
X	conversion	–	–
X	number of moles	lbmol	mol
X	quality of steam	–	–
y	mass fraction in extract	–	–
y	vapor mole fraction	–	–
y	volume fraction	–	–
Y	expansion coefficient	–	–
Y	ordinate in the Eckert's flooding curve	lbm-ft-cP$^{0.1}$/lbf	m/s
z	mole fraction of the vapor stream in a still	–	–
z	datum elevation	ft	m

Symbols

α	relative volatility	–	–
β	diameter ratio	–	–
ϵ	emissivity	–	–
ϵ	fraction of area available for vapor flow	–	–
ϵ	particle porosity	–	–
ϵ	roughness of pipe	ft	m
η	efficiency	–	–
θ	hydraulic residence time	min	min
θ	retention time	min	min
λ	latent heat	Btu/lbm	kJ/kg
μ	viscosity	lbm/ft-sec	cP
ν	kinematic viscosity	ft^2/sec	m^2/s
ρ	density	lbm/ft^3	kg/m^3
σ	Stefan-Boltzmann constant	Btu/ft^2-hr-°R^4	W/m^2·K^4
σ	stoichiometric coefficient	–	–
σ	surface tension	lbf/ft	N/m
τ	residence time	min	min
υ	specific volume	ft^3/lbm	m^3/kg, L/kg
ϕ	relative humidity	–	–
ω	specific humidity	lbm water/lbm dry air	g water/g dry air

Subscripts

a	activation, air, ambient, angle valve, or available		h	hydrocarbon phase
abs	absolute		HC	hydrocarbon layer
ac	acetone		HP	high pressure
anhy	anhydride		hw	hot water
ave	average		i	internal, component, or stage number
B	bottoms or globe valve		ifb	insulating firebrick
c	capital, combustion, compressor constant, concentration, or convection		ind	indirect
			ins	instrument
			inv	investment
ch	cold hydrocarbon		j	stage number
con	conical		L	liquid phase
cond	condensation		lb	labor
cr	capital recovery		lbmn	labor and maintenance
cw	cold water		lm	log mean of two values
cyl	cylindrical		LP	low pressure
d	square-edged orifice		lv	liquid vapor
D	distillate		m	maintenance, maximum, mixture, or motor
di	dirty inside		M	condition mixture or mixing point
dil	dilution		min	minimum
dir	direct		mlb	maintenance labor
do	dirty outside		mmt	maintenance materials
e	elbow, elevation, or equivalent		mwb	mineral-wool brick
elc	electricity		n	stage net normal
emul	emulsion		N	stage
eq	equilibrium or equipment		nf	no flooding
f	flooding, flow, formation, or friction		o	external, initial condition, orifice, or outer external
F	feed			
fb	firebrick		oi	other indirect
fdi	fouling based on the internal diameter		olb	operating labor
fdo	fouling based on the external diameter		op	operating
fg	flue gas		over	overhead
g	gas phase (mass) or gate valve		p	packing, product, or pump
G	gas phase (molar)		peq	purchased equipment

r	radiation or refrigeration	
R	reactor, rectification section, or recycle	
rb	refractory brick	
rec	recovered	
relc	required electricity	
rxn	reaction	
s	isentropic, saturated, saturation, side, static, stream, or surface	
S	side, soluble, solvent, steam, stripping section, or suspended solids	
sb	scrubber	
sfr	superficial flow rate	
slb	supervisory labor	
sol	solution	
t	total or turbine	
u	utility	
U	uncontrolled	
v	valve	
v	velocity	
V	vapor phase	
vap	vaporization or vapor stream	
w	waste, water, or wet bulb	
W	batch still	
wg	waste gas	
X	liquid phase or stripping section	
Y	gas phase	

Superscripts

0	pure species, reference condition, or standard state
$*$	equilibrium

Problems

MASS/ENERGY BALANCES AND THERMODYNAMICS

PROBLEM 1

A perfectly insulated covered tank maintained at atmospheric pressure is 10 ft high and 16.5 ft in diameter. The tank is half-filled with water at 80°F. The density of water at this temperature is 62.21 lbm/ft^3. The heat capacity of liquid water is 1.0 Btu/lbm-°F. The heat of vaporization of water at 212°F is 970.3 Btu/lbm. The heat capacity of the water vapor is 0.48 Btu/lbm-°F. The enthalpy change of the water as a result of a temperature change from 80°F to 350°F is most nearly

(A) 13×10^6 Btu
(B) 69×10^6 Btu
(C) 78×10^6 Btu
(D) 160×10^6 Btu

Hint: Calculate the enthalpy change as the temperature of the water increases from 80°F to 212°F. Then calculate the enthalpy change as the water evaporates completely and the enthalpy change as the temperature of the water vapor increases to 350°F.

PROBLEM 2

Dry air and vinyl chloride (C_2H_3Cl) are combined to produce a saturated mixture at 14.3 lbf/in^2 and 77°F. At this temperature, the vapor pressure of vinyl chloride is 5.77 lbf/in^2. The average molecular weight of vinyl chloride and dry air are 62.50 lbm/lbmol and 28.96 lbm/lbmol, respectively. Assume the ideal gas law applies. The mass ratio of vinyl chloride to dry air in the saturated mixture is most nearly

(A) 0.676
(B) 1.06
(C) 1.40
(D) 1.46

Hint: Calculate the mole fraction of vinyl chloride using the vapor pressure of vinyl chloride and the total pressure. Use the molecular weights of vinyl chloride and air to convert the mole fraction to the mass ratio.

PROBLEM 3

Dry air and vinyl chloride (C_2H_3Cl) are mixed to produce a saturated mixture at 3739 mm Hg and 20°C. At this temperature, the vapor pressure of vinyl chloride is 2580 mm Hg. The average molecular weights of vinyl chloride and dry air are 62.50 g/mol and 28.96 g/mol, respectively. Assume the ideal gas law applies. In 100 cm^3 of saturated mixture, the mass of vinyl chloride is most nearly

(A) 0.0088 g
(B) 0.41 g
(C) 0.88 g
(D) 13 g

Hint: Use the ideal gas law to calculate the density of vinyl chloride.

PROBLEM 4

A process stream consisting of 22% component M (by weight) and 4.0% component P (by weight) in water at 48,000 lbm/hr is produced by mixing two streams, S1 and S2. Stream S1 is produced in mixer 1 by mixing two streams, J1 and J2. Stream J1 is 99% M (by weight) in water, and stream J2 is 95% P (by weight) in water. There is no M or P in stream S2.

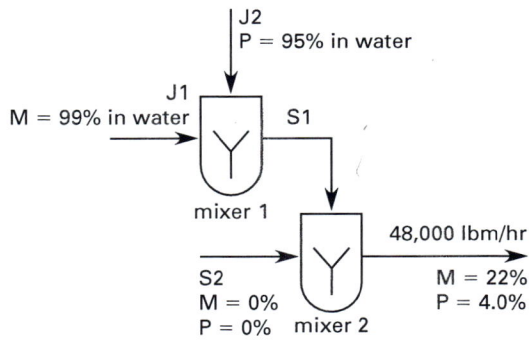

The percentage of component P in stream S1 is most nearly

(A) 2.0%
(B) 15%
(C) 46%
(D) 83%

Hint: The mass flow rate of component P in stream S1 must equal the mass flow rate of P in the desired product because S1 is the sole source of P in the desired product. Performing a mass balance around mixer 1 yields the mass flow rate of S1.

PROBLEM 5

A process stream consisting of 20% M (by weight), 4.0% P (by weight), and 24% R (by weight) in water at 36,000 lbm/hr is produced by mixing two streams, S1 and S2. Stream S1 is produced in mixer 1 by mixing two streams, J1 and J2. Stream J1 is 95% M (by weight) in water, and stream J2 is 95% P (by weight) in water. Stream S2 is produced in mixer 2 by mixing two streams, J3 and J4. Stream J3 consists of 48% R (by weight) in water, and stream J4 consists of pure water.

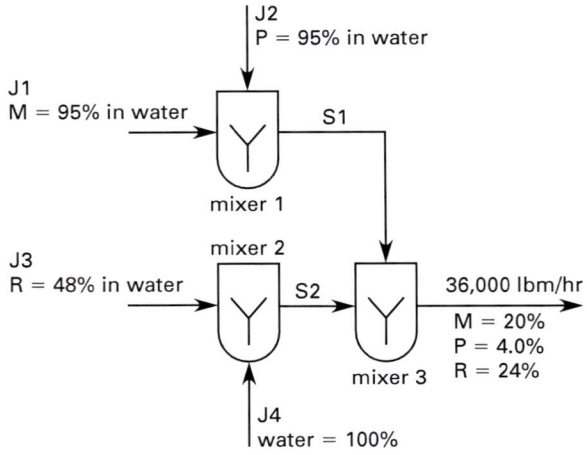

The mass flow rate of water in J4 is most nearly

(A) 8900 lbm/hr
(B) 9800 lbm/hr
(C) 28,000 lbm/hr
(D) 29,000 lbm/hr

Hint: Perform mass balances around the equipment to solve for the flow rates and compositions of process streams.

PROBLEM 6

The irreversible reaction M → 2N + P is carried out in a catalytic reactor that operates at steady state. The feed to mixer 1 consists of pure M. The feed enters mixer 1 at a molar flow rate of 100 lbmol/hr. The reactor effluent enters a separator. In the reactor, 98% of M (per mole) is consumed. A portion of the separator effluent is recycled to mixer 1. The product stream contains components M, N, and P. The recycle stream contains components M and N. The product stream contains 0.6% (per mole) of the total component M in lbmol/hr entering the separator. The portion of component N in the recycle stream is 2% (per mole) of the component N entering the separator.

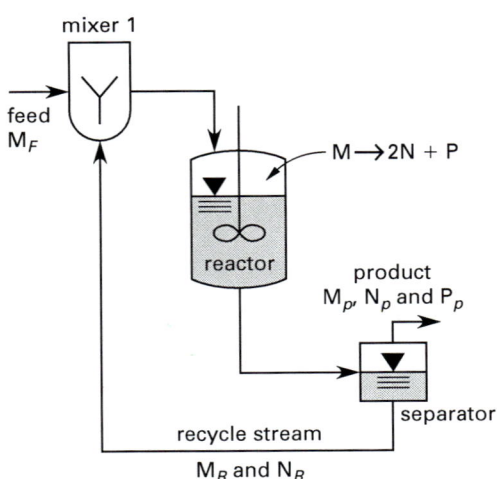

The molar flow rate of N in the recycle stream is most nearly

(A) 0.01 lbmol/hr
(B) 2 lbmol/hr
(C) 4 lbmol/hr
(D) 200 lbmol/hr

Hint: Calculate the molar flow rates of M and N in the recycle stream by performing a mass balance of M and N around mixer 1, the reactor, and the separator.

PROBLEM 7

1 mol of carbon monoxide (CO) reacts with 3 mol of hydrogen (H_2) at 100°C and 1 atm to give methane (CH_4) and water (H_2O). For the reaction $CO(g) + 3H_2(g) \rightarrow CH_4(g) + H_2O(g)$, the standard heats of formation and the free energies of formation at 25°C are

component	ΔH^0_{298} (cal/mol)	ΔG^0_{298} (cal/mol)
$CO(g)$	$-26\,398$	$-32\,762$
$H_2(g)$	0	0
$CH_4(g)$	$-17\,798$	$-12\,052$
$H_2O(g)$	$-57\,757$	$-54\,593$

The numerical value of the equilibrium constant for the reaction at 100°C is most nearly

(A) 2×10^3
(B) 4×10^{17}
(C) 1×10^{27}
(D) 4×10^{65}

Hint: Use the stoichiometric coefficients from the balanced reaction to calculate the Gibbs free energy and the enthalpy change for the reaction at 25°C. Then, calculate the Gibbs free energy at 100°C and the equilibrium constant.

PROBLEM 8

Liquid methylhydrazine (CH_6N_2) reacts with liquid dinitrogen tetroxide (N_2O_4) to give gaseous carbon dioxide (CO_2), gaseous water (H_2O), and gaseous nitrogen (N_2)

at 25°C and 1 atm. For the reaction, the standard heats of formation and the entropies of formation at 25°C are

component	ΔH^0_{298} (kJ/mol)	S^0_{298} (J/mol·K)
$CH_6N_2(l)$	54.14	165.94
$N_2O_4(l)$	9.08	304.38
$CO_2(g)$	−393.51	213.785
$H_2O(g)$	−241.83	188.835
$N_2(g)$	0	191.56

The Gibbs free energy for the reaction at 25°C is most nearly

(A) −5500 kJ/mol
(B) −4800 kJ/mol
(C) −4000 kJ/mol
(D) −800 kJ/mol

Hint: Be careful to get the units correct. The enthalpy data are in kilojoules per mole while the entropy data are in joules per mole Kelvin. Use the stoichiometric coefficients from the balanced reaction to calculate the enthalpy change and entropy change for the reaction at 25°C.

PROBLEM 9

A gas stream containing 46.2% (by mole) butane, 53.1% (by mole) air, and 0.7% (by mole) water is to be processed to recover the butane. The butane that has a molecular weight of 58.08 lbm/lbmol is to be separated from the remaining vapors by cooling the stream and thus condensing out the butane. The feed flows into the process at a rate of 300 ft³/min at a temperature of 77°F and a pressure of 14.7 lbf/in². The required removal efficiency is 85%. The mass flow rate of the butane recovered is most nearly

(A) 17.5 lbm/hr
(B) 18.0 lbm/hr
(C) 1050 lbm/hr
(D) 1230 lbm/hr

Hint: Calculate the molar volume and the flow rate of butane in the inlet stream. Next, calculate the butane molar flow rate in the outlet stream and the molar flow rate of the butane condensed. Then, calculate the mass flow rate of butane recovered.

PROBLEM 10

A vent stream consisting of acetone, air, and a negligible amount of moisture flows at 200 ft³/min at standard conditions. The acetone is to be separated from the remaining vapors by condensation. This process requires a refrigerated condenser system. The inlet volume fraction of acetone in the vent stream is 0.385. The required removal efficiency is 90%. The relationship between the temperature in degrees Celsius and the vapor pressure of the acetone in mm Hg is

$$\log p_{ac,\text{mm Hg}} = 7.117 - \frac{1210.595 K}{T + 229.664 K}$$

The temperature required to achieve the desired removal efficiency is most nearly

(A) −78°F
(B) −8.2°F
(C) 17°F
(D) 130°F

Hint: Calculate the partial pressure of acetone at the outlet of the condenser. Solve the Antoine equation to calculate the temperature necessary to condense the required amount of acetone.

PROBLEM 11

The volume of water in a continuously stirred tank reactor is 5400 ft³. In this tank, water is treated with a chloramine. The flow rate through the reactor is 1,188,300 ft³/day. The chloramine is continuously injected at an inflow concentration of 0.000936 lbm/ft³, beginning at time zero. The decomposition of the chloramine in the water follows a first-order reaction. The concentration of the chloramine in the reactor is 0 lbm/ft³ at time zero. The reaction rate is slow, so it may be assumed to be negligible. Assume the reactor volume is constant. The output concentration, C, of the chloramine as a function of the time, t, is

(A) $C = 0.000936 \left(1 - e^{-t/6.54 \text{ min}}\right)$
(B) $C = 0.000936 \left(1 - e^{-t/0.00454 \text{ min}}\right)$
(C) $C = 0.000936 \left(1 + e^{-t/6.54 \text{ min}}\right)$
(D) $C = -0.000936 \left(1 - e^{-t/6.54 \text{ min}}\right)$

Hint: Evaluate the rate of reaction at exit conditions.

PROBLEM 12

An adiabatic humidifier system equipped with a humidifier, a liquid water tank, and a sprinkler system operates at the conditions given.

The heat of vaporization of water at 212°F is 970.3 Btu/lbm. The heat capacity of water vapor is 0.48 Btu/lbm-°F. The enthalpies of the saturated gas at 350°F and 800°F are 62.3 Btu/lbm and 230.1 Btu/lbm, respectively. The water droplets from the sprinkler system are used to cool a 4000 ft³/min (measured at 68°F and 407 in H_2O) dry gas stream entering the top of the humidifier. The water tank has a volume of 2036 ft³ and is 70% filled. Assume the ideal gas law applies.

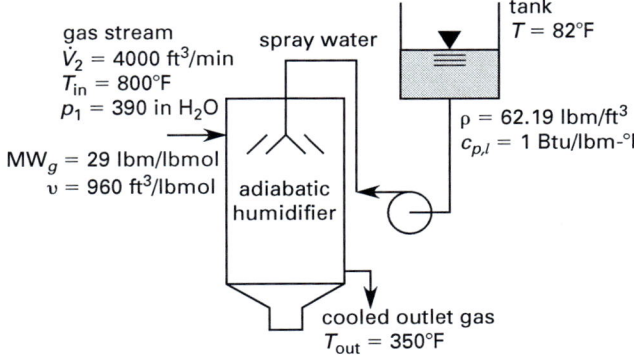

The water in the tank is sufficient for a period of most nearly

(A) 34 hr
(B) 49 hr
(C) 92 hr
(D) 2300 hr

Hint: This process is an adiabatic energy transfer from the gas stream entering the humidifier to the water in the water tank as the water evaporates.

PROBLEM 13

Two pure process streams are fed to a perfectly mixed tank kept at 77°F. One stream consists of pure compound M. The other stream is water (H_2O). The solution in the tank is produced at a flow rate of 28,800 lbm/hr of a 15% (by weight) solution. The molecular weight of component M is 42.39 lbm/lbmol and that of H_2O is 18.015 lbm/lbmol. In the evaporator, 18,000 lbm/hr of H_2O are removed. The enthalpy of solution of M in H_2O is given at different concentrations in the following table.

concentration (lbmol H_2O/lbmol M)	enthalpy of solution of M at 77°F (Btu/lbmol)
17.31	−33,174
13.30	−32,036
8.85	−28,435
5.70	−25,812
3.53	−22,046

The rate of the heat produced in the tank is most nearly

(A) 0.0 Btu/hr
(B) 2.3×10^6 Btu/hr
(C) 2.6×10^6 Btu/hr
(D) 3.3×10^6 Btu/hr

Hint: Consider the process of preparing a 15% solution of M in water. For this solution, the heat released is the enthalpy change for the corresponding mixing process.

PROBLEM 14

A shift reactor receives three streams. Stream 1 is fed to the reactor at a rate of 100 lbmol/hr and consists of 76% (by mole) nitrogen (N_2), 18% (by mole) carbon monoxide (CO), and carbon dioxide (CO_2). Stream 2 consists of 48% (by mole) hydrogen (H_2) and CO. Stream 3 consists of steam. The outlet from the reactor consists of two streams—streams 4 and 5. Stream 4 is pure CO_2. Stream 5 consists of H_2 at 75% (by mole) and N_2. The ratio of the molar flow rate of stream 1 to steam stream 3 is most nearly

(A) 0.33
(B) 0.48
(C) 0.79
(D) 1.3

Hint: Perform mole balances of N_2, CO, H_2, and CO_2 around the reactor.

PROBLEM 15

A solution consisting of 15% M (by mass) and water (H_2O) at 25°C with a negligible boiling point elevation is concentrated in a single-effect evaporator to 40% M (by mass). The solution is fed at a flow rate of 4.0 kg/s. The evaporator is operating at atmospheric pressure. The steam to the evaporator is at 39.2 atm with a latent heat of 1715.3 kJ/kg. The product leaves the evaporator at 153°C. The enthalpy of saturated H_2O at 25°C is 104.89 kJ/kg and that of saturated steam at 153°C is 2750 kJ/kg. At a pressure of 1 atm, the specific heat of the solution produced in the evaporator is 2.72 kJ/kg·°C. The molecular weight of M is 42.39 g/mol. At 1 atm, the heat of solution of M to form 1 mol of an aqueous solution is −23.267 kJ/mol of M for the solution produced by the evaporator and −33.810 kJ/mol of M for the feed. The mass flow rate of saturated steam is most nearly

(A) 0.39 kg/s
(B) 3.7 kg/s
(C) 4.2 kg/s
(D) 4.6 kg/s

Hint: The heat of solution is the enthalpy change per mole of solute (M) of the process of dissolving 1 mol of solute in the solvent, H_2O, at 25°C and 1 bar. An enthalpy balance around the evaporator gives the heat the steam must provide to the solution.

PROBLEM 16

The waste gas preheater of an incinerator receives a waste gas at 110°F. The flue gas from the combustion chamber of the incinerator is used to preheat the incoming feed stream and combustion air. The heat capacities of the gases on both sides of the preheater are approximately the same. The incinerator has an operating temperature of 1600°F and a fractional energy recovery of 0.75. The maximum energy recoverable would be the decrease in sensible heat of the flue gas if it were cooled to the temperature of the incoming waste gas. Assume the mass flow rates of gases on both sides of the preheater are approximately the same.

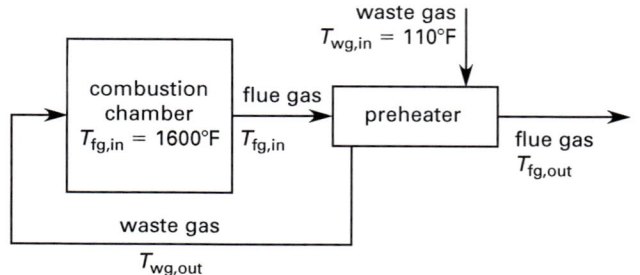

The outlet flue gas temperature at the preheater is most nearly

(A) 110°F
(B) 320°F
(C) 480°F
(D) 2900°F

Hint: Perform mass and energy balances around the preheater. Apply the definition of the fraction energy recovery.

PROBLEM 17

A combustion chamber receives a feed that consists of a gaseous waste containing benzene 1000 ppm (by volume), chloroform 1000 ppm (by volume), and 77°F air at 1 atm. The molecular weight of the waste gas is 28.97 lbm/lbmol. At 77°F, the volumetric heat of combustion is 3616 Btu/ft^3 for benzene and 705 Btu/ft^3 for chloroform. The heat of combustion of the waste gas stream is most nearly

(A) −58 Btu/lbm
(B) −39 Btu/lbm
(C) −4.3 Btu/lbm
(D) 58 Btu/lbm

Hint: Calculate the mole fractions of benzene and chloroform. Calculate the heat of combustion by adding the heats of combustion of chloroform and benzene.

PROBLEM 18

In a steady-flow adiabatic combustor, gaseous n-pentane (C_5H_{12}) is mixed with air and burned to CO_2 and CO using 100% excess air. The gases enter the chamber at 25°C and leave at 827°C. Assume air is 79% (by mole) nitrogen and 21% (by mole) oxygen, and assume constant average heat capacities are acceptable over the range. Average heat capacities and enthalpies of formation are given in the following table.

component	$\overline{C}_p$ (cal/mol·K)	ΔH_f at 25°C (kcal/mol)
C_5H_{12}	53.08	−30.086
O_2	7.82	0
CO_2	7.47	−94.0518
CO	11.61	−26.4157
$H_2O(g)$	9.02	−57.798
N_2	7.39	0

The percentage of n-pentane burned to CO_2 is most nearly

(A) 6.0%
(B) 9.0%
(C) 13%
(D) 17%

Hint: Determine the amount of each component present in the chamber in terms of the amount of C_5H_{12} burned to CO_2. Perform and enthalpy balance around the adiabatic combuster.

PROBLEM 19

An insulated heat exchanger is carrying saturated steam at a pressure of 24 bar. An oxygen stream at a pressure of 1.5 bar and flowing at the rate of 150 kmol/h is to be heated from 25°C to 220°C using the heat exchanger. The average heat capacity of oxygen is constant over the range at 30.07 kJ/kmol·°C. The steam consumption is most nearly

(A) 0.080 kg/h
(B) 3.0 kg/h
(C) 400 kg/h
(D) 480 kg/h

Hint: Use the steam tables to determine the condensation enthalpy of the steam. Perform an energy balance between the oxygen stream and the saturated steam.

PROBLEM 20

The volume of water in a continuously stirred tank reactor (CSTR) is kept constant at 5600 ft^3. In this tank, water is treated with a chloride. The liquid flow rate through the reactor is 1.2×10^6 ft^3/day. A reactive substance, S, that has a first-order decay rate of 0.25 hr^{-1} also flows in the CSTR. The input concentration of S has been 0.0125 lbmol/ft^3 for a long time, and the system has reached steady state. At a given time, the input concentration of S is increased to 0.0250 lbmol/ft^3. The steady-state concentration of S in the CSTR before the concentration of S increased is most nearly

(A) 0.0045 lbmol/ft^3
(B) 0.012 lbmol/ft^3
(C) 0.019 lbmol/ft^3
(D) 0.025 lbmol/ft^3

Hint: Calculate the hydraulic residence time in the tank and the first-order decay rate.

PROBLEM 21

A feed of 12,000 lbm/hr consists of 90% (by mass) of a solution of sugars in water and 10% (by mass) of suspended solids. The feed is first split off, resulting in two streams—streams 2 and 3. The fraction of stream 3 is 12% (by mass) of the feed. Stream 2 is fed to an evaporator operated below atmospheric pressure. The 12% (by mass) stream produced in the splitter is mixed with the outlet stream from the evaporator. The evaporator produces a concentrate containing 85% (by mass) suspended solids.

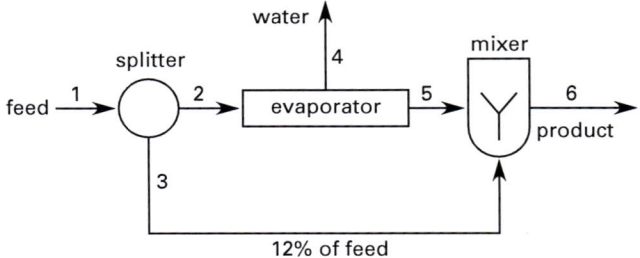

The mass fraction of the suspended solids in the final product is most nearly

(A) 0.34
(B) 0.45
(C) 0.48
(D) 0.95

Hint: Perform total mass and suspended solids balances around the splitter. Perform suspended solids and water balances around the evaporator. Perform total mass and suspended solids balances around the mixer.

PROBLEM 22

A flow splitter (S1) divides a liquid stream flowing at 1000 lbmol/hr into three streams as shown. The stream flow rates are regulated so that the flow rate in stream 2 is two times the flow rate in stream 3 and the flow rate in stream 3 is one-third of the flow rate in stream 4. The second splitter (S2) divides stream 3 into two streams flowing at equal flow rates. Stream 6 is combined with stream 4 in mixer 1 to produce stream 7.

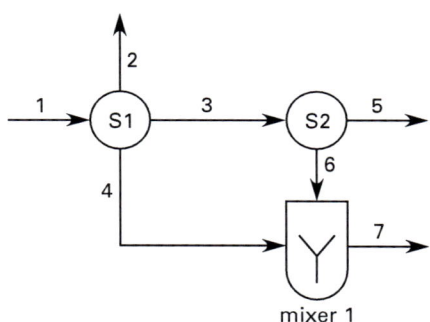

The flow in stream 7 is most nearly

(A) 420 lbmol/hr
(B) 560 lbmol/hr
(C) 580 lbmol/hr
(D) 630 lbmol/hr

Hint: Perform a mole balance around each flow splitter.

PROBLEM 23

Isobutane (iC_4H_{10}) reacts at a rate of 191 kmol/h with butene (C_4H_8) in an emulsion with concentrated sulfuric acid (H_2SO_4) to produce iso-octane (C_8H_{18}). The feed to the process contains 25% (by mole) iC_4H_{10}, 25% (by mole) C_4H_8, and 50% (by mole) n-butane (C_4H_{10}). The feed combines with three separate recycle streams as shown, and the combined stream enters the reactor. The reactor effluent is an emulsion that enters a splitter that separates the emulsion into two streams. One stream from the splitter is recycled to the reactor inlet and the other is fed to the decanter where it is separated into two liquid phases. The stream to the decanter contains 1300 kmol/h iC_4H_{10}. The bottom layer from the decanter is an H_2SO_4-rich stream that contains 2 kg of H_2SO_4 per kilogram of hydrocarbon. This stream is recycled to the process. The upper layer from the decanter is a hydrocarbons layer that is fed to a distillation column. The overhead from the column contains C_8H_{18} and C_4H_{10}. The distillation bottom is pure iC_4H_{10} and is recycled to the reactor. Assume C_4H_{10} is chemically inert.

component	molecular weight, MW (kg/kmol)	reactor inlet, F (kmol/h)
C_4H_{10}	58.12	17270
iC_4H_{10}	58.12	43391
C_4H_8	56.10	217
C_8H_{18}	114.22	8418

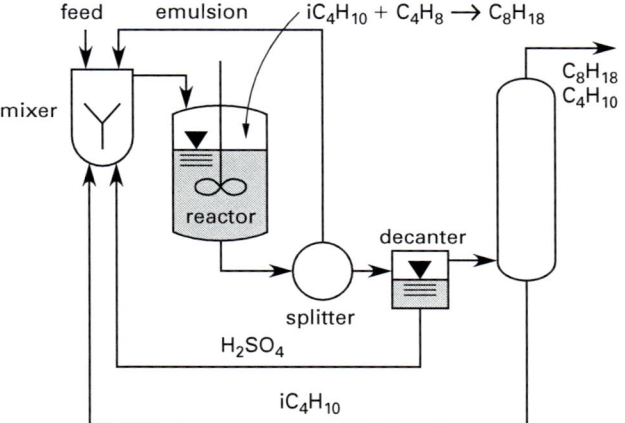

The mass flow rate of H_2SO_4 in the emulsion recycle stream is most nearly

(A) 2.1×10^5 kg/h
(B) 2.7×10^5 kg/h
(C) 4.5×10^5 kg/h
(D) 8.9×10^6 kg/h

Hint: Calculate the mass flow rate of the hydrocarbons going into the reactor. Perform a mole balance of iC_4H_{10} around the reactor and distillation column.

PROBLEM 24

Ammonia (NH_3) and oxygen (O_2) react to give nitrogen monoxide (NO) and water (H_2O) according to the reaction $4NH_3 + 5O_2 \leftrightarrow 4NO + 6H_2O$. The conversion of the limiting reactant is 85% (by mole). The feed consists of an equimolar mixture of ammonia and oxygen fed at the rate of 120 lbmol/hr. The output rate of the product stream from the reactor is most nearly

(A) 8.00 lbmol/hr
(B) 130 lbmol/hr
(C) 132 lbmol/hr
(D) 135 lbmol/hr

Hint: Determine the limiting reactant. Use the stoichiometry of the reaction to calculate the output rate of the product stream.

PROBLEM 25

1 mol of carbon dioxide (CO_2) reacts with 4 mol of hydrogen (H_2) in a reactor at 400°C and 1 atm to give methane (CH_4) and water (H_2O). For the reaction $CO_2(g) + 4H_2(g) \rightarrow CH_4(g) + 2H_2O(g)$, the standard heats of formation, the free energies of formation at 25°C (298K), and the average heat capacities are

component	ΔG^0_{298} (J/mol)	ΔH^0_{298} (kJ/mol)	$\overline{C}_p$ (J/mol·K)
$CO_2(g)$	$-394,359$	-393.51	52.55
$H_2(g)$	0	0	30.12
$CH_4(g)$	$-50,460$	-74.87	50.81
$H_2O(g)$	$-228,572$	-241.83	34.71

The number of moles of $CO_2(g)$ converted to $CH_4(g)$ at 400°C is most nearly

(A) 0.82 mol
(B) 0.84 mol
(C) 0.98 mol
(D) 1.0 mol

Hint: Be careful to get the units correct. Note that the Gibbs free energy data are in Joules per mole and the enthalpy data are in kilojoules per mole. Calculate the Gibbs free energy at 400°C and the equilibrium constant.

PROBLEM 26

A steam power plant operates on a reheat Rankine cycle at a mass flow rate of 1 kg/s. The steam enters the high-pressure turbine at 12 MPa and 612°C and is condensed to saturated liquid in the condenser at a pressure of 12 kPa. The saturated liquid is then pumped to the boiler. The temperature of the fluid at state 5 is the same as the temperature at state 3. The moisture content in the low-pressure turbine is not to exceed 8%. Assume all devices operate with 100% efficiency.

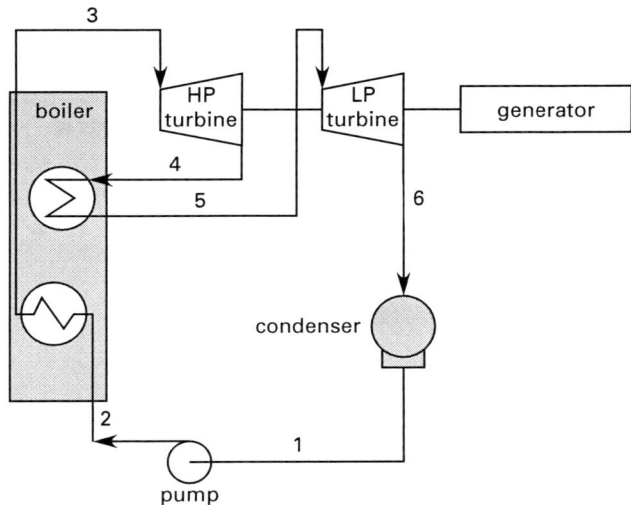

The thermal efficiency of the Rankine cycle operating at these conditions is most nearly

(A) 44%
(B) 51%
(C) 64%
(D) 100%

Hint: Calculate the maximum work in each turbine. For maximum work the process is reversible.

PROBLEM 27

A stream consisting of saturated steam at 356°F with a bulk velocity of 130 ft/sec and a flow rate of 250 lbm/hr goes through a boiler that adds heat to the stream at the rate of 300 Btu/lbm. The superheated steam subsequently expands through a back-pressure turbine where it develops 60 hp of shaft work and exits through a diffuser with a velocity of 1 ft/sec. The elevation change between the inlet and the outlet of the system is 220 ft. Assume negligible pressure drop due to friction.

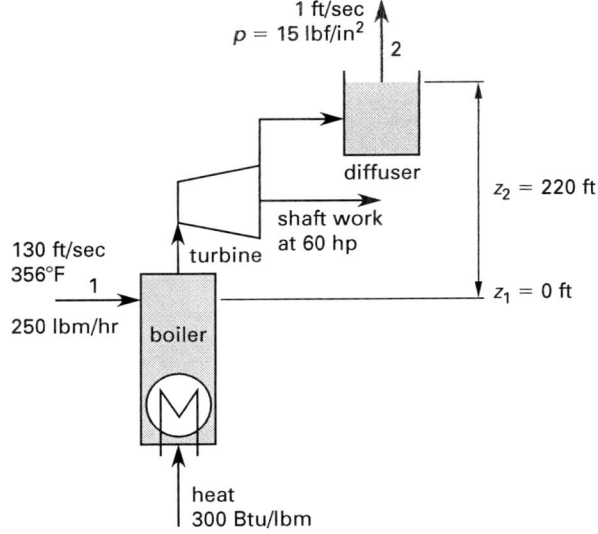

The quality of the steam at the outlet of the diffuser is most nearly

(A) 0
(B) 0.1
(C) 0.7
(D) 1

Hint: Perform an energy balance to calculate the enthalpy at outlet conditions. The diffuser decreases the kinetic energy of a fluid by increasing the pressure.

PROBLEM 28

The compressor of a simple-cycle gas turbine operates with an efficiency of 87% while the turbine operates with an efficiency of 90%. The compressor pressure ratio is 5 and the pressure drop in the combustion chamber is 6%. The turbine inlet temperature is 1500°F. Air enters the compressor at 70°F and 1 atm. The air has an isentropic exponent of 1.40 and a constant heat capacity of 0.24 Btu/lbm-°R. The combustion gas has a heat capacity of 0.2744 Btu/lbm-°R and an isentropic exponent of 1.333.

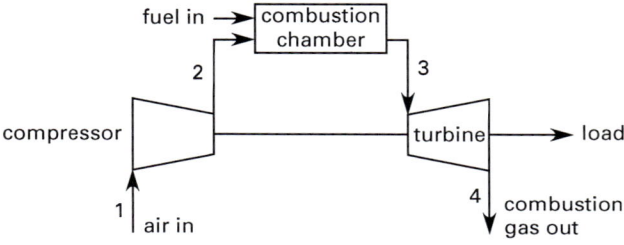

The thermal efficiency of this simple-cycle gas turbine is most nearly

(A) 0.22
(B) 0.23
(C) 0.29
(D) 0.81

Hint: Consider the cycle consisting of an isentropic compression of the gas, a constant-pressure heat addition, an isentropic expansion of the gas through the turbine, and an isobaric closure of the cycle back to the starting point.

PROBLEM 29

An ideal gas flowing at a rate of 15 mol/s at a temperature of 298K and at 120 kPa is to be compressed to 1600 kPa using a three-stage isentropic compressor. The compressor system has an efficiency of 90% and is driven by an electric motor with 70% efficiency. The compression ratio is equal for each stage. Assume a constant outlet temperature of 298K for the intercoolers. The compression power required by the three-stage compressor is most nearly

(A) 110 kW
(B) 170 kW
(C) 680 kW
(D) 790 kW

Hint: Calculate the ideal work and then account for compression and electric inefficiencies.

PROBLEM 30

A vent stream at a temperature of 77°F and a pressure of 14.7 lbf/in², consisting of benzene, air, and a negligible amount of moisture, flows at 200 ft³/min. Benzene has a molecular weight of 78 lbm/lbmol and is to be separated from the remaining vapors through saturation followed by condensation in a refrigerated condenser system. The condenser is a shell-and-tube countercurrent heat exchanger, and the inlet volume fraction of benzene is 0.35. The required removal efficiency is 92%. The coolant is ethylene glycol. The recovered benzene stream is free of moisture, and the ideal gas law applies. The mass flow rate of benzene recovered is most nearly

(A) 10 lbm/hr
(B) 13 lbm/hr
(C) 770 lbm/hr
(D) 2200 lbm/hr

Hint: Calculate the volume of benzene using the ideal gas law. Calculate the molar flow rate of benzene and convert this flow rate to mass flow rate to obtain the answer.

FLUIDS

PROBLEM 31

A centrifugal pump supplies water at 180°F at a rate of 100 gal/min from a tank under vacuum to a tank vented to the atmosphere. The pump suction line is a 10 ft length of 2 in schedule-40 pipe. The pipe's inner diameter is 2.067 in and its internal area is 0.0233 ft². The suction line contains one long-radius screwed elbow and a valve. The elbow resistance coefficient is 0.4, and the valve resistance coefficient is 5.6. Under these conditions, the head loss due to friction in the suction line is 17.4 ft for each 100 ft of pipe.

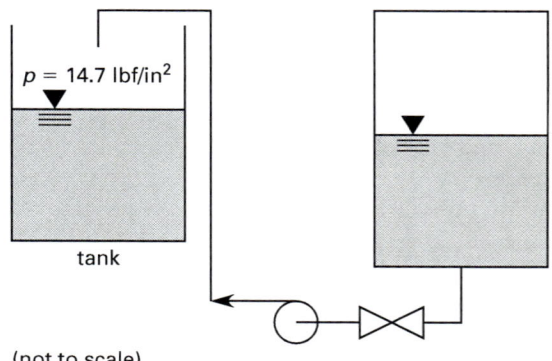

The friction loss in the suction line pipe and fitting is most nearly

(A) 2.3 ft
(B) 9.7 ft
(C) 10 ft
(D) 280 ft

Hint: Calculate the velocity head and the head loss through the valve and elbow.

PROBLEM 32

A centrifugal pump supplies water at 170°F at a rate of 86 gal/min from a tank under a 20 in Hg vacuum to a tank vented to the atmosphere. The liquid level above the pump centerline is 16 ft. The pump suction line is a 12 ft length of 2 in schedule-40 pipe. The internal area of the pipe is 0.0233 ft^2. The suction line contains one long-radius elbow with a resistance coefficient of 0.4 and a valve with a resistance coefficient of 5.6. 170°F water has a vapor pressure of 6.0 lbf/in^2 and a density of 62.4 lbm/ft^3. Under these conditions, the head loss due to friction in the suction line is 16.4 ft for each 100 ft of pipe. Assume the specific gravity of mercury is 13.61.

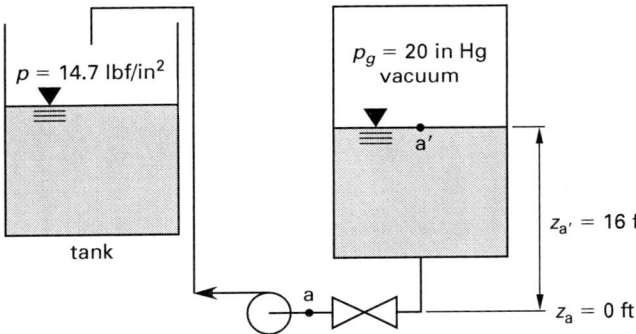

(not to scale)

The net positive suction head available is most nearly

(A) 5.0 ft
(B) 22 ft
(C) 33 ft
(D) 50 ft

Hint: Calculate the atmospheric pressure head, the vapor pressure head, the static suction head, and the head loss due to friction.

PROBLEM 33

The following figure shows the separator for a refinery main fractionator overhead system. Mixed-phase material enters from the main fractionator overhead condensers. The feed separates into two layers. The vapors produced in the separator are routed to the wet gas compressor. Water is pumped from the bottom of the separator and sent to a water treatment unit. The hydrocarbon layer is sent to further product separation units. An internal standpipe in the separator transfers a hydrocarbon-layer-only stream to the hydrocarbon layer pumps (not shown). The suction-line loss due to friction is 2.8 ft. The minimum water layer level above the datum line is 2 ft. The minimum water layer thickness in the separator is 1 ft, and the minimum hydrocarbon layer thickness is 3.5 ft. At 120°F, the hydrocarbon layer density is 43.68 lbm/ft^3 and the operating pressure above the liquid is 21.6 lbf/in^2. 120°F water has a vapor pressure of 1.70 lbf/in^2 and a density of 62.4 lbm/ft^3. To minimize the pressure drop between the liquid level upstream and the pump suction, the pump inlet velocity head is extremely low and can be ignored. Assume the bottom of the separator is 1 ft above the datum line.

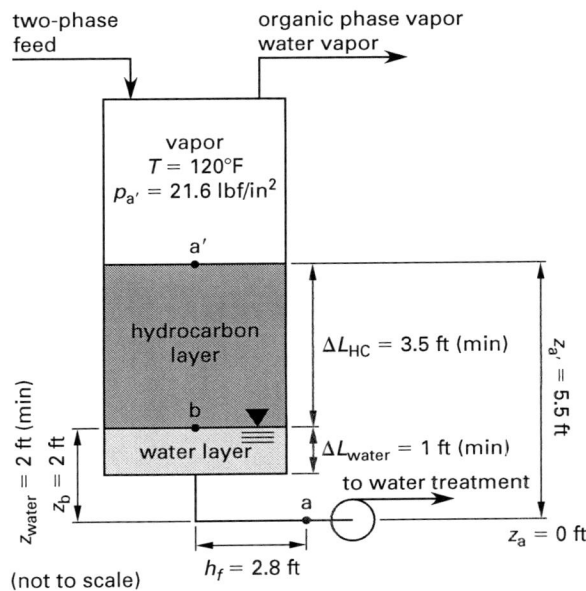

The net positive suction head available is most nearly

(A) 43 ft
(B) 45 ft
(C) 48 ft
(D) 53 ft

Hint: Calculate the head loss due to pressure and the static head loss.

PROBLEM 34

The vessel shown contains an organic phase and an aqueous phase. The vapors formed in the vessel at 120°F and 21.6 lbf/in^2 go to a compressor. The aqueous phase is pumped from the bottom of the vessel and sent to a water treatment unit. An internal standpipe (not shown) in the vessel allows pumping of the organic phase stream to a purification unit. The loss due to friction in the suction line is 3.8 ft. The minimum water level is 2 ft and the minimum organic phase layer thickness is 5.5 ft. At 120°F, the organic phase specific gravity is 0.70. The vapor pressure of water at 120°F is 1.70 lbf/in². To minimize the pressure drop between the liquid level upstream and the pump suction, the water pump inlet velocity head is extremely low and can be ignored. Assume the bottom of the vessel is 1 ft above the datum line.

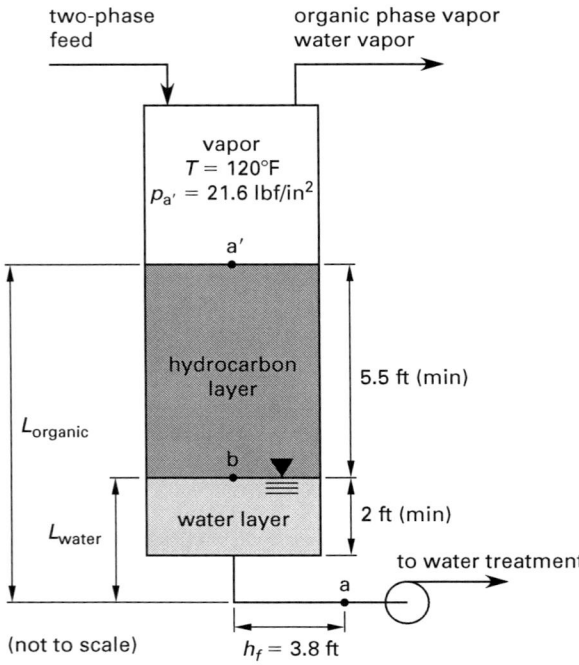

(not to scale)

With all other conditions unchanged, a decrease in organic phase specific gravity would

(A) increase the water pump pressure head
(B) increase the velocity head
(C) decrease the static head of the water pump
(D) increase the loss in pressure due to frictional losses in flow

Hint: Consider the static head produced by the fluid between the liquid level and the pump suction centerline.

PROBLEM 35

A pump supplies 330 gal/min of a fluid with a density of 61.8 lbm/ft³ and a viscosity of 0.9 cP to a tank vented to the atmosphere. The fluid is pumped at 77°F through a 4 in schedule-40 steel pipe with an equivalent length of pump discharge line of 483 ft. The pipe roughness is 0.000144 ft. The head loss due to friction is most nearly

(A) 2.3 ft
(B) 3.4 ft
(C) 28 ft
(D) 56 ft

Hint: Calculate the Reynolds number and the friction factor.

PROBLEM 36

A fluid with a density of 61.8 lbm/ft³ is pumped at a flow rate of 345 gal/min to a tank vented to the atmosphere. The fluid is pumped at 77°F through a 2 in schedule-40 steel pipe with an equivalent length of pump discharge line of 500 ft. The internal diameter of the pipe is 2.067 in. The Darcy friction factor is 0.029213. The line has a globe control valve in the full open position with an equivalent length of 110 ft. The expression used to calculate the flow coefficient of the control valve is most nearly

(A) $C_v = \dfrac{345 \, \frac{\text{gal}}{\text{min}}}{\sqrt{758 \, \frac{\text{lbf}}{\text{in}^2}}}$

(B) $C_v = \dfrac{345 \, \frac{\text{gal}}{\text{min}}}{\sqrt{622 \, \frac{\text{lbf}}{\text{in}^2}}}$

(C) $C_v = \dfrac{345 \, \frac{\text{gal}}{\text{min}}}{\sqrt{137 \, \frac{\text{lbf}}{\text{in}^2}}}$

(D) $C_v = \dfrac{345 \, \frac{\text{gal}}{\text{min}}}{\sqrt{135 \, \frac{\text{lbf}}{\text{in}^2}}}$

Hint: Calculate the velocity of the fluid and the pressure drop head through the valve.

PROBLEM 37

A fluid is pumped at 77°F through a 4 in schedule-40 steel pipe to a tank vented to the atmosphere. The pump supplies 330 gal/min. The internal diameter of the pipe is 4.026 in. The line has a globe control valve and an orifice plate. The orifice plate has a coefficient of 0.60 and a permanent pressure drop of 54%. The ratio of the orifice diameter to the pipe diameter is 0.68. The head loss through the orifice is most nearly

(A) 2.1 ft
(B) 5.9 ft
(C) 11 ft
(D) 20 ft

Hint: Calculate the velocity of the fluid and the scale reading of the orifice plate. Then calculate the permanent pressure drop through the orifice.

PROBLEM 38

A liquid with a density of 56.13 lbm/ft³ and viscosity of 470 cP is flowing at a rate of 800 gal/min through a 6 in schedule-40 pipe. The internal diameter of the pipe is 6.065 in. In the piping system shown, point 2 is 20 ft higher than point 1. The pressure at point 2 is

15 lbf/in². The resistance coefficients for the two valves and the elbow are given in the following table.

description	symbol	value
elbow	K_e	0.30
gate valve	K_g	0.12
angle valve	K_a	2.25

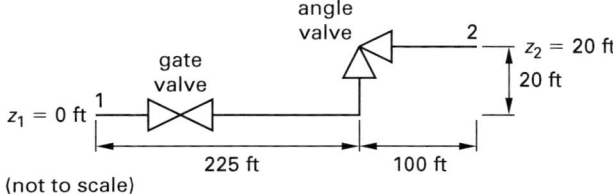

(not to scale)

The pressure at point 1 is most nearly

(A) 20 lbf/in²
(B) 26 lbf/in²
(C) 50 lbf/in²
(D) 1100 lbf/in²

Hint: Calculate the Reynolds number, the velocity, and the pressure difference between points 1 and 2.

PROBLEM 39

A tank is filled with a liquid that has a density of 123.6 lbm/ft³. The discharge piping is 1 in schedule-40 steel pipe. The internal diameter of the pipe is 1.049 in and the pipe is 54 ft long. The roughness of the pipe is 0.00015 ft. The pressure drop due to friction is 26.35 lbf/in². The velocity of the liquid in the pipe is 10 ft/sec. The viscosity of the fluid is most nearly

(A) 0.012 cP
(B) 13 cP
(C) 18 cP
(D) 80 cP

Hint: Calculate the friction factor and the Reynolds number. From the Reynolds number, determine if the flow is laminar or turbulent.

PROBLEM 40

A centrifugal pump with a pump head of 107 ft is used to supply water at 40°F to a vented tank from a nearby lake. The water surface in the storage tank is 75 ft higher than the lake surface. The equivalent length of pipe connecting the lake, pump, and tank, including all fittings and entrance and exit effects, is 200 ft. The diameter of the pipe is 2.067 in. For this pipe, the roughness is 0.00015 ft. Assume that water at 40°F has a density of 62.43 lbm/ft³ and a viscosity of 1.42 cP. The flow rate at which water at 40°F can be delivered to the tank is most nearly

(A) 90 gal/min
(B) 170 gal/min
(C) 180 gal/min
(D) 230 gal/min

Hint: Use Bernoulli and Darcy equations together with the Reynolds number to calculate the pump head and the friction factor in terms of the flow rate. Solve these equations simultaneously.

PROBLEM 41

A centrifugal pump with an efficiency of 58% is used to supply 26 402.39 kg/h water at 20°C to a vented tank from a nearby lake. The water surface in the vented storage tank is 3 m higher than the lake surface. A 50 m equivalent length of 2 in nominal-size schedule-40 steel pipe connects the lake, the pump, and the tank. This equivalent length includes all fittings and entrance and exit effects. The pipe has an internal diameter of 0.0525 m and an internal area of 0.002165 m². The viscosity of water at 20°C is 1 cP. Assume a roughness value of 0.0457 mm. The density of water at 20°C is 998.2 kg/m³. The minimum power required to drive the pump is most nearly

(A) 18 W
(B) 190 W
(C) 1800 W
(D) 6.5×10^6 W

Hint: Express the friction head loss in terms of the friction factor and the velocity of the fluid. Express the velocity of the fluid in terms of the volumetric flow rate, and express the friction factor in terms of the velocity. Solve these equations simultaneously.

PROBLEM 42

T.W. Cochran postulated that the friction head loss, h_f, in feet, is equal to the square of the volumetric flow rate, Q, in cubic feet per second times the resistance coefficient, K, dimensionless; that is, $h_f = KQ^2$. In the piping network shown, the flow rates at points R and U are 350 gal/min. The head loss between points R and U is 67.4 ft. The surface roughness is 0.0002 ft for all piping. The pipe network is in the horizontal plane. The Darcy friction coefficient is given by $f = \left(2\log(\epsilon/3.7D)\right)^{-2}$.

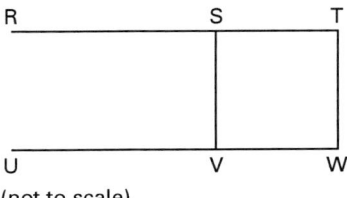

(not to scale)

The following table shows the equivalent pipe lengths and the diameters of the piping network.

line	L (ft)	d (in)
RS	750	5
SV	450	5
ST	450	5
TW	450	5
WV	450	5
VU	750	4

The flow rate in the SV line is most nearly

(A) 0.50 gal/min
(B) 3.7 gal/min
(C) 30 gal/min
(D) 220 gal/min

Hint: Calculate the friction and resistance coefficients in all the lines. Then use the Bernoulli equation to calculate the volumetric flow rate in the SV line.

PROBLEM 43

T.W. Cochran postulated that the friction head loss, h_f, in feet, is equal to the square of the volumetric flow rate, Q, in cubic feet per second times the resistance coefficient, K, dimensionless; that is, $h_f = KQ^2$. The simplified piping network shown is carrying a fluid at 68°F (density = 56.6 lbm/ft^3, viscosity = 0.00027 lbm/sec-ft). All the pipe lengths are given as equivalent pipe lengths. The flow rate at points A and E is 400 gal/min. The head loss between points A and E is 49.79 ft. A surface roughness of 0.0002 ft is assigned for all piping, which is assumed to be in a horizontal plane. The friction factor is given by $f = (-2\log(\epsilon/3.7D))^{-2}$.

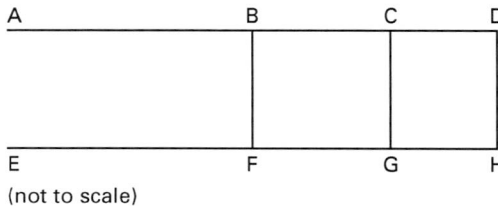

(not to scale)

The lengths and diameters of the lines are

line	L (ft)	d (in)	D (ft)	resistance coefficient
AB	800	5	0.41667	K_{AB}
EF	800	5	0.41667	K_{EF}
BF	400	4	0.33333	K_{BF}
BC	450	5	0.41667	K_{BC}
FG	450	5	0.41667	K_{FG}
CD	350	5	0.41667	K_{CD}
GH	350	5	0.41667	K_{GH}
CG	400	5	0.41667	K_{CG}
DH	400	5	0.41667	K_{DH}

The Reynolds number in the BF line is most nearly

(A) 122,000
(B) 300,400
(C) 338,000
(D) 570,900

Hint: Calculate the resistance coefficient, the flow rate, the area, and the velocity of the fluid in line BF.

PROBLEM 44

Flue gas at 320°F is fed to a smoke stack through a 3 in schedule-40 pipeline. The internal diameter of the pipe is 3.068 in. To determine the gas flow, a horizontal venturi meter having a throat of 1.25 in is installed. For the existing steady-state conditions, a mercury manometer indicates a pressure differential of 20 in. The inlet pressure is 14.5 lbf/in^2 and the gas contains 0.5 mol% carbon dioxide (CO_2). The average molecular weight of the gas is 29.3 lbm/lbmol and that of CO_2 is 44 lbm/lbmol. The ratio of specific heat capacities of the gas is 1.41. The discharge coefficient of the venturi meter is a straight line when plotted against the ratio of the diameters of the throat and the pipe. This straight line has a slope of 0.0123 and an intercept of 0.9878. Assume an isentropic flow of an ideal gas and that all measurements are made well downstream of any flow disturbance. The mass emission rate of CO_2 from the stack is most nearly

(A) 0.008 lbm/hr
(B) 8 lbm/hr
(C) 30 lbm/hr
(D) 50 lbm/hr

Hint: Calculate the discharge coefficient of the venturi meter, the velocity of the flue gas, and the flue gas volumetric mass flow rate.

PROBLEM 45

A square-edged orifice is to be used to measure the flow rate of ammonia at 32°F and 60 lbf/in^2 in a 12 in schedule-40S (stainless steel) pipe. The pipe has an internal diameter of 12.00 in and an internal area of 0.7854 ft^2. The maximum ammonia flow is 6000 ft^3/min, and the maximum instrument pressure reading across the orifice is 127 in H_2O. The square-edged orifice coefficient is 0.595. At 32°F, ammonia has a viscosity of 2.205×10^{-4} lbm/ft-sec and a density of 0.0482 lbm/ft^3. The ratio of specific heats is 1.3. The diameter of the orifice is most nearly

(A) 1.7 in
(B) 4.4 in
(C) 5.7 in
(D) 7.1 in

Hint: Express the flow and the expansion coefficients of the orifice in terms of the ratio of the diameter of the orifice to the internal diameter of the pipe. Express the volumetric flow rate in terms of the flow and the expansion coefficients of the orifice. Solve simultaneously.

HEAT TRANSFER

PROBLEM 46

A heat exchanger having one pass on the shell side and two passes on the tube side is used to cool 1000 kg/h of a hydrocarbon mixture. The hydrocarbon mixture has a specific heat capacity of 2.594 kJ/kg·°C. The hydrocarbon enters the exchanger at 232°C and leaves at 176°C. The hydrocarbon mixture is being cooled by water flowing in the shell. The water enters at 140°C and leaves at 154°C. The heat capacity of the water is 2.427 kJ/kg·°C. The shell-side mass flow rate is most nearly

(A) 270 kg/h
(B) 2300 kg/h
(C) 3700 kg/h
(D) 4300 kg/h

Hint: Perform an energy balance on the heat exchanger. The heat gained by the hydrocarbon in the tube side is equal to the heat lost by the water in the shell of the exchanger.

PROBLEM 47

A heat exchanger having one pass on the shell side and two passes on the tube side is used to cool 1000 kg/h of a butane stream. The butane stream has a specific heat capacity of 2.594 kJ/kg·°C. This stream enters the exchanger at 232°C and leaves at 176°C. The butane stream is cooled by water flowing in the shell. The water enters the shell at 140°C and leaves at 154°C. The heat capacity of the water is 2.427 kJ/kg·°C. The overall heat-transfer coefficient of the exchanger based on the inside area is 410 kJ/h·m²·°C, and the tube diameter is 2.54 cm. The overall heat-transfer coefficient of the exchanger based on the outside area is 205 kJ/m²·h·°C, and the shell diameter is 5.08 cm. The tube surface area is most nearly

(A) 6.79 m²
(B) 12.7 m²
(C) 13.1 m²
(D) 13.6 m²

Hint: Calculate the logarithmic mean temperature difference and correct it using the F_T chart correction factor correlation. Use the equation for rate of heat exchange to calculate the area of heat exchange.

PROBLEM 48

A heat exchanger having an overall heat-transfer coefficient, based on the outside area, of 800 Btu/ft²-hr-°F and a tube diameter of 0.277 ft is used to cool a hexanes mixture. The heat exchanger is one pass on the shell side and two passes on the tube side. The heat exchanger is used to cool 50,000 lbm/hr of a hexanes mixture with a specific heat capacity of 0.82 Btu/lbm-°F. The hexanes mixture enters the heat exchanger at 225°F and leaves at 174°F. The cooling fluid is water flowing in the shell entering the heat exchanger at 140°F and leaving at 150°F. The overall heat-transfer coefficient based on the inside area is 20 Btu/ft²-hr-°F. For a 10-tube hexagonal array, the length of the tube is most nearly

(A) 5.8 ft
(B) 6.0 ft
(C) 7.3 ft
(D) 240 ft

Hint: Calculate the corrected logarithmic mean temperature difference (LMTD) using the correction factor correlation and the area of heat exchange. Use the surface area and the number of tubes to calculate the length of each tube.

PROBLEM 49

A process stream C1 is fed to heat exchanger 1 with an overall heat-transfer coefficient of 800 W/m²·K. A second process stream, C2, is fed to heat exchanger 2 with an overall heat-transfer coefficient of 1200 W/m²·K. The mass flow rate of process stream C1 is 1 kg/min, and the mass flow rate of process stream C2 is 2 kg/min. Streams C1 and C2 need to be heated in their respective heat exchangers from 300K to 480K prior to reaction. The reactor effluent then needs to be cooled from 500K to 400K in heat exchanger 3 with an overall heat-transfer coefficient of 800 W/m²·K. The heat capacities of all streams are constant at 60 kJ/kg·K. Assume a capital recovery factor of 0.2 over a 10 year period, a heat exchanger capital cost of $2714 times the area in square meters, an operation of 8000 h/yr, and the following cost and usage data.

stream	characteristic	symbol	value
hot water	heat duty	HD_{hw}	450 kW
	cost	C_{hw}	8×10^{-7}/kJ
steam	heat duty	HD_S	540 kW
	cost	C_S	5×10^{-6}/kJ
cold water	heat duty	HD_{cw}	300 kW
	cost	C_{cw}	3×10^{-7}/kJ

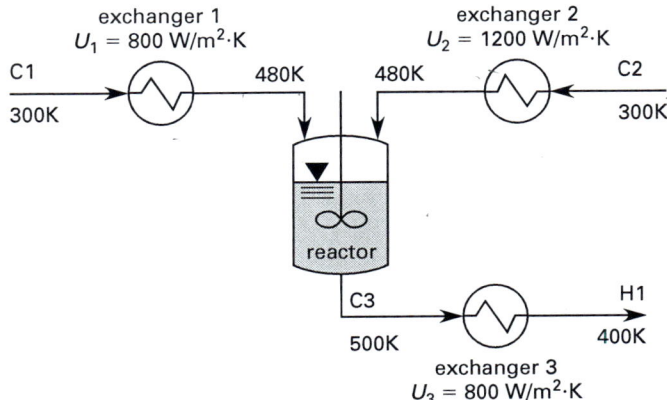

The annual capital and utility cost is most nearly

(A) $84,000/yr
(B) $91,000/yr
(C) $92,000/yr
(D) $94,000/yr

Hint: Calculate the required heating area for each of the heat exchangers. Calculate the amount of steam, hot water, and cooling water required.

PROBLEM 50

A horizontal pipe with an outside diameter of 8.625 in carries saturated steam. The line is not insulated and is 20 ft long. The temperature of ambient air is 68°F. The rate of heat loss per foot of pipe length is 2115.5 Btu/hr. The combined heat-transfer coefficient for convection and radiation is 2.4 Btu/ft^2-hr-°F. Assume the steam-film resistance, pipe-wall resistance, and thickness are negligible. The temperature at the surface of the pipe is most nearly

(A) 68°F
(B) 88°F
(C) 320°F
(D) 460°F

Hint: The total heat transferred is related to the external area of the pipe, the temperature at the surface of the pipe, and the air temperature. Perform the calculation based on a one-foot length of pipe.

PROBLEM 51

A double-effect evaporator is used to concentrate a citric acid solution. The feed enters the second effect at 70°F, and saturated steam enters the first effect at 220°F. The temperature in the final condenser is 110°F. The overall heat-transfer coefficients in the evaporators are 450 Btu/hr-ft^2-°F and 315 Btu/hr-ft^2-°F for the first and second effects, respectively. The heating areas for the two effects are the same. The solution has a negligible elevation in boiling point and a negligible heat of dilution. Radiation losses may be neglected. The temperature in the first effect is most nearly

(A) 110°F
(B) 160°F
(C) 170°F
(D) 220°F

Hint: Perform a heat-transfer balance around the two evaporators in terms of the unknown temperature in the first effect. Solve the equation for the temperature in the first effect.

PROBLEM 52

A heat exchanger condenses 2.7×10^6 lbm/hr of propane at 150°F and 200 lbf/in^2 using water. Propane has a heat capacity of 0.39 Btu/lbm-°F and a condensation temperature of 105.2°F. The water enters the heat exchanger at 86°F and leaves at 96°F. The overall heat-transfer coefficient on the basis of the total outside heat-transfer area is 95.9 Btu/ft^2-hr-°F. The latent heat of propane at 150°F and 200 lbf/in^2 is 138.1 Btu/lbm. Assume the condensing side is approximately isothermal. The heat-transfer area of the heat exchanger is most nearly

(A) 3.6×10^4 ft^2
(B) 2.9×10^5 ft^2
(C) 3.2×10^5 ft^2
(D) 4.4×10^5 ft^2

Hint: Calculate the total duty of the heat exchanger.

PROBLEM 53

A shell-and-tube heat exchanger is used to cool an organic fluid in a countercurrent fashion. Water enters the shell at a mass flow rate of 1000 lbm/hr. The shell fluid enters at a temperature of 80°F and leaves at 140°F. The organic fluid enters at a flow rate of 353 lbm/hr and a temperature of 450°F. The heat capacities of the shell-side and tube-side fluids are 1 Btu/lbm-°F and 0.85 Btu/lbm-°F, respectively. Heat losses to the air from the outer pipe and the pipe-wall resistance can be neglected. The logarithmic mean temperature difference is most nearly

(A) 120°F
(B) 210°F
(C) 230°F
(D) 350°F

Hint: Calculate the temperature of the organic liquid as it leaves the tube side. Application of the definition will give the logarithmic mean temperature difference.

PROBLEM 54

A shell-and-tube heat exchanger is used to cool an organic fluid using water in the shell in a countercurrent fashion. Heat losses to the air from the outer pipe and the pipe-wall resistance can be neglected. The shell fluid enters at 80°F and leaves at 140°F. The organic fluid is cooled from 450°F to 250°F. The shell-side heat-transfer coefficient is 800 Btu/ft^2-hr-°F. The tube-side heat-transfer coefficient is 250 Btu/ft^2-hr-°F. The shell-side and tube-side fouling factors are 0.001 ft^2-hr-°F/Btu and 0.002 ft^2-hr-°F/Btu, respectively. The viscosity correction factor is $(\mu/\mu_w)^{0.14}$, where μ_w is the viscosity of the fluid at the temperature of the wall. The data available are listed in the following table.

shell-side viscosity (cP)	shell-side temperature (°F)	tube-side viscosity (cP)	tube-side temperature (°F)
0.65	110	0.90	155
0.50	139	0.83	167
0.47	146	0.60	208
0.43	155	0.45	234
0.37	167	0.30	300
–	–	0.14	350

The overall fouled-clean equivalent heat-transfer coefficient, corrected for viscosity, is most nearly

(A) 73 Btu/ft²-hr-°F
(B) 170 Btu/ft²-hr-°F
(C) 190 Btu/ft²-hr-°F
(D) 280 Btu/ft²-hr-°F

Hint: Determine the wall temperature on the tube side and shell side. Calculate the viscosity correction factor and the overall heat-transfer coefficient.

PROBLEM 55

The insulation system of a furnace wall is constructed using refractory brick in the inner wall followed by firebrick, mineral wool brick, and insulating firebrick in the outer wall. The inner and outer surface temperatures of the insulated wall are 1100°C and 30°C, respectively. The thicknesses and thermal conductivities of the wall materials are given in the following table.

material	symbol	thickness, L (m)	thermal conductivity, k (W/m·K)
refractory brick	rb	0.3	45
firebrick	fb	0.6	0.11
mineral wool brick	mwb	0.5	0.3
insulating firebrick	ifb	0.3	0.02

The interface temperature between the firebrick and the mineral wool brick is most nearly

(A) 100°C
(B) 300°C
(C) 800°C
(D) 1400°C

Hint: Calculate the thermal resistances of a 1 m² area of each of the four components of the wall. Then determine the overall heat-loss rate from 1 m² of wall.

PROBLEM 56

A steel pipe with an outside diameter of 8.625 in carrying steam is insulated with 1.5 in of kapok fibers surrounded by 1 in of rubber foam. The ambient air is at 70°F. The emissivity of the surface of the pipe is 0.85. The convection and radiation heat-transfer coefficients are uniform along the exposed surface of the pipe. The temperature at the surface of the insulated pipe is 156°F. Assume steady-state conditions and that the steam-film and pipe-wall resistances are negligible. For a horizontal pipe, the convection heat-transfer coefficient is

$$h_c = \left(0.5 \, \frac{\text{Btu-in}^{0.25}}{\text{ft}^2\text{-hr-°R}^{1.25}}\right) \left(\frac{T_s - T_a}{d_0}\right)^{0.25}$$

The combined heat-transfer coefficient is most nearly

(A) 0.80 Btu/ft²-hr-°F
(B) 1.9 Btu/ft²-hr-°F
(C) 2.6 Btu/ft²-hr-°F
(D) 4.3 Btu/ft²-hr-°F

Hint: Calculate the convection and radiation heat-transfer coefficients.

PROBLEM 57

A system consisting of a refrigeration unit, a condenser, and a recovery tank condenses a volatile organic compound. The refrigeration unit required has a cost of $36,523. The condenser cost is estimated at $12,615. The recovery tank cost can be estimated at $3940. The purchased equipment cost is 1.18 times the equipment cost. The total capital investment is 1.74 times the purchased equipment cost. The total capital investment cost is most nearly

(A) $63,000
(B) $92,000
(C) $100,000
(D) $110,000

Hint: Calculate the equipment cost, the purchased equipment cost, and the total capital investment.

PROBLEM 58

A refrigeration unit with a capacity of 14.1 tons of refrigeration operates 8 hr/day, 5 days/wk. The operating cost is based on 30 min/day, 5 days/wk, 52 wk/yr at $16.07/hr. The supervisory labor is computed at 15% of the operating cost. The maintenance costs are the labor and the material costs. The maintenance labor and maintenance material costs are equal and are based on 30 min/shift at $17.50/hr. Each shift is 8 hr. Power is consumed at a rate of 4.4 kW/ton and a cost of $0.0461/kW/hr with a compressor efficiency of 85%. The total direct annual cost is most nearly

(A) $9400/yr
(B) $11,700/yr
(C) $13,600/yr
(D) $14,000/yr

Hint: Calculate the operating cost, the supervisor cost, the maintenance labor, the maintenance material, and the electricity cost.

PROBLEM 59

A facility operates 8 hr/day, 5 days/wk, 52 wk/yr. The total capital investment is estimated at $110,873. The operating cost is estimated at $2230/yr. The supervisory labor is estimated at $334/yr. The operating labor and maintenance materials costs are each estimated at $2314/yr. The indirect annual costs are comprised of 4% of the total capital investment, capital recovery cost, and overhead. The annual capital recovery cost is 0.1098 times the total capital investment. The overhead cost is 61% of total labor and maintenance material. The general and administrative, property tax, and insurance costs are 4% of the total capital investment. The total indirect annual cost is most nearly

(A) $16,400/yr
(B) $16,600/yr
(C) $19,400/yr
(D) $20,800/yr

Hint: Calculate the total capital investment, the capital recovery cost, and the overhead cost.

PROBLEM 60

A refrigeration unit recovers 600 lbm/hr of acetone. The acetone recovered resale value is $0.10/lbm. The total capital investment cost is estimated at $103,687. The operating labor cost is estimated at $2452/yr. The supervisory labor is computed at 15% of the operating labor cost. The maintenance labor and maintenance material costs are equal and are estimated at $2545/yr each. The electricity cost is estimated at $7078/yr. The indirect annual costs are comprised of 4% of the total capital investment, capital recovery cost, and overhead. The overhead cost is 61% of the total labor and maintenance material cost. The total labor cost is estimated at $2820/yr. The annual capital recovery cost is 10.98% of the total capital investment. The general and administrative, property tax, and insurance costs are 4% of the total capital investment. The facility operates 8 hr/day, 5 days/wk, 52 wk/yr. If the condensed acetone can be directly reused or sold without further treatment, the total savings is most nearly

(A) $89,500/yr
(B) $104,000/yr
(C) $110,000/yr
(D) $125,000/yr

Hint: Subtract the total direct annual cost and the total annual indirect cost from the acetone recovered resale value.

MASS TRANSFER

PROBLEM 61

A mixture of hydrocarbons is flash-distilled at 86.5°F and 150 lbf/in². The molar flow rate of the feed is 100 lbmol/hr. The molar flow rate of the vapor stream in the feed is 39 lbmol/hr. The feed compositions, z_i, and the vapor-liquid constants, K_i, for the various hydrocarbons at 86.5°F and 150 lbf/in² are given in the following table.

hydrocarbon	z_i	K_i
ethane	0.002	3.89
propane	0.040	1.43
propene	0.340	1.21
butane	0.152	0.39
pentane	0.340	0.26
hexane	0.126	0.14

The mole fraction of pentane in the liquid phase after flash distillation is most nearly

(A) 0.1
(B) 0.4
(C) 0.5
(D) 0.6

Hint: Calculate the mole fraction of pentane in the vapor phase. Use this value to calculate the mole fraction of pentane in the liquid phase.

PROBLEM 62

A fluid with a kinematic viscosity of 1.384×10^{-5} m²/s, a density of 1.3 kg/m³, and a porosity of 0.431 travels a column 1 m long filled with particles 6.35×10^{-3} m in diameter. The flow velocity is 2.15 m/s. The pressure drop through the bed is most nearly

(A) 0.0 Pa
(B) 2300 Pa
(C) 5700 Pa
(D) 12 000 Pa

Hint: Calculate the Reynolds number and use the Ergun equation.

PROBLEM 63

A double-effect evaporator is used to concentrate a salicylic acid solution. The feed enters the second effect at 70°F, and saturated steam enters the first effect at 220°F. The temperature in the final condenser is 110°F. The overall heat-transfer coefficients in the effects are 450 Btu/hr-ft²-°F and 315 Btu/hr-ft²-°F for the first and second effects, respectively. The first effect has 1.8 ft² of heat-transfer area while the second effect has 3.3 ft². The solution has a negligible elevation in boiling point and a negligible heat of dilution. Assume that the radiation losses may be neglected and that the rate of heat transferred is approximately equal in both effects. The temperature in the first effect is most nearly

(A) 140°F
(B) 160°F
(C) 170°F
(D) 190°F

Hint: Perform a heat-transfer balance around the two evaporators in terms of the unknown temperature in the first effect. Solve the equation for the temperature in the first effect.

PROBLEM 64

The waste gas flow rate entering an absorber is 21,377 standard ft^3/min (22,288 actual ft^3/min). The pollutant in the waste gas is hydrochloric acid (HCl). There are 1871 ft^3 of HCL entering the absorber in the waste gas per 10^6 ft^3 of waste gas. The pollutant removal efficiency is 99% (by mole). The absorbant is a caustic solution with the same physical properties as water. The actual liquid-to-gas ratio is 1.5 times the minimum liquid-to-gas ratio. The maximum HCl concentration in the liquid phase, in equilibrium with the HCl entering the column in the gas phase, is 0.16 lbmol HCl/lbmol HCl-free solvent. The HCl concentration exiting the absorber in the liquid stream is most nearly

(A) 0.00110 lbmol HCl/lbmol solvent
(B) 0.106 lbmol HCl/lbmol solvent
(C) 0.108 lbmol HCl/lbmol solvent
(D) 0.160 lbmol HCl/lbmol solvent

Hint: Calculate the actual liquid-to-gas ratio and the HCL concentrations in the gas phase.

PROBLEM 65

A countercurrent absorber is used to absorb 95% (by mole) of the hydrofluoric acid (HF) in a waste gas entering the absorber at a flow rate of 23,000 ft^3/min. The absorber uses a caustic solution as the solvent. The waste gas treated contains 0.18% (by mole) of HF and has a density of 0.0709 lbm/ft^3. The molecular weight of waste gas is 29 lbm/lbmol. The maximum HF concentration in the liquid stream in equilibrium with HF entering the column in the gas stream is at 0.15 lbmol HF per lbmol of HF-free solvent. The actual liquid-to-gas ratio is 1.6 times the minimum. Under these operating conditions, Henry's law applies and the equilibrium line equation, in terms of the mole fractions, is $y = 0.00104x$. The absorption factor is most nearly

(A) 0.018
(B) 11
(C) 18
(D) 53,000

Hint: Calculate the actual liquid-to-gas ratio. Then calculate the liquid and gas total molar flow rates.

PROBLEM 66

A stream consisting of 28% (by mole) of an organic compound in hexane is flashed adiabatically. The flash distillation is performed in a still that holds a nominal 100 lbmol. The still was originally charged with 85 lbmol. To prevent loss of agitation, the still is not distilled below a level of 15 lbmol. In this distillation, hexane has an average relative volatility of 3.2. The final mole fraction of the organic compound in the liquid phase is most nearly

(A) 5.0×10^{-4}
(B) 2.0×10^{-3}
(C) 3.0×10^{-3}
(D) 2.0×10^{-2}

Hint: Apply the Rayleigh equation to a batch distillation binary system.

PROBLEM 67

A liquid feed contains 40% (by mole) methyl ethyl ketone (MEK) and 60% (by mole) water at 760 mm Hg. Assume that Raoult's law applies. The Antoine constants, with pressure in mm Hg and temperature in the absolute scale, are

component	A	B (K)	C (K)
MEK	14.2173	2831.82	-57.3831
water	18.3036	3816.44	-46.13

The relative volatility of MEK with respect to water is most nearly

(A) 0.23
(B) 1.3
(C) 1.5
(D) 3.8

Hint: From the Antoine constants, determine the temperature of the liquid mixture at the total pressure.

PROBLEM 68

A flash distillation is to be performed in a still in two batches. The first batch is to be charged with 87 mol of a binary ideal solution consisting of 74% (by mole) of solvent A and 26% (by mole) of solvent B. After the vaporization occurs in each batch, the liquid that remains in the still is 23% (by mole) of the charge. The second batch is charged with pure solvent B to a total of 87 mol. Ignoring the effect of any other organic compounds in the still on the equilibrium, the solvents have a constant relative volatility of 2.7. The total number of moles of solvent B distilled after the second distillation is most nearly

(A) 50 mol
(B) 60 mol
(C) 70 mol
(D) 90 mol

Hint: Apply the Rayleigh equation to the solution.

PROBLEM 69

An oil-sand mixture that is 25% (by mass) oil and 75% (by mass) sand is to be extracted or leached with 75 tons/day of naphtha in a countercurrent extractor. The feed consists of 100 tons/day of mixture. The final extract (overflow) produced contains 35% (by mass) oil and 65% (by mass) naphtha, and the underflow from each unit consists of 32% (by mass) solution and 68% (by mass) sand. The overall efficiency of the extraction is 80% (by mass). Assume the solvent is miscible with the oil in all proportions and the extractor has reached equilibrium conditions in each stage. Assume there is no sand in the overflow. The number of stages required to effect the desired separation of oil from sand is

(A) 3
(B) 4
(C) 5
(D) 6

Hint: Perform an overall mass balance around the system and a mass balance around the nth stage.

PROBLEM 70

A certain process chamber requires 1.2×10^6 ft^3/hr air at 72°F and 50% relative humidity. This air is to be obtained by mixing fresh air at 32°F and 15% relative humidity with 600,000 ft^3/hr of recycled air at 72°F and 55% relative humidity. The mixture of fresh and recycled air is to be first heated, then humidified adiabatically to the desired specific humidity, and finally reheated to the desired temperature. The relative humidity of the air leaving the humidifier is 75%. For the humidifier, the heat-transfer coefficient between the air and the surface of the water multiplied by the mass transfer area is expected to be 85 Btu/ft^3-hr-°F. The size of humidifier required is most nearly

(A) 188 ft^3
(B) 372 ft^3
(C) 377 ft^3
(D) 1120 ft^3

Hint: Perform a mass balance to calculate the mass flow rate of the fresh air. Determine the specific humidity and temperature of the streams. Determine the average heat capacity of the air in the humidifier.

PROBLEM 71

A copper ore containing 10.3% (by mass) copper sulfate, 85.4% (by mass) inert, and 4.3% (by mass) water is to be extracted with pure water in a countercurrent extractor. The daily feed consists of 281 tons. The final extract produced contains 10% (by mass) copper sulfate and 90% (by mass) water. The underflow from each stage consists of 66.7% (by mass) solution and 33.3% (by mass) inert. The process is to recover 92% of the copper sulfate from the ore. Assume the extractor has reached equilibrium conditions in each stage. The minimum number of stages required to effect the desired separation of copper sulfate from the inert is

(A) 3
(B) 5
(C) 6
(D) 7

Hint: Perform a mass balance around the entire system and around the first and n^{th} steps.

PROBLEM 72

Propane (C_3H_8) is burned in a steady-flow adiabatic combustor to produce CO_2 and CO using controlled excess air. All gases enter the combustion chamber at 25°C. The chamber is at atmospheric pressure. The number of moles of CO_2 produced per mole of C_3H_8 burnt is 0.76. The gases leave the chamber at 845°C. The average heat capacities over the temperature range and the enthalpies of formation at 25°C are given in the following table.

component	ΔH_f (cal/mol)	$\overline{C}_p$ (cal/mol·K)
O_2	0	7.83
CO_2	$-94\,051.8$	11.65
CO	$-26\,415.7$	7.49
$H_2O(g)$	$-57\,798$	9.05
N_2	0	7.40
C_3H_8	$-26\,600$	33.14

The excess air used is most nearly

(A) 112%
(B) 129%
(C) 132%
(D) 212%

Hint: Balance the combustion reaction. Then perform an energy balance around the combustion chamber to calculate the number of moles of oxygen used during the combustion.

PROBLEM 73

A vent stream consisting of methanol, air, and a negligible amount of moisture flows at 200 ft^3/min (measured at a temperature of 77°F and a pressure of 14.7 lbf/in^2). The methanol is to be separated from the remaining vapors through saturation followed by condensation in a refrigerated condenser system. The inlet stream temperature is 85°F, and the inlet volume fraction of methanol is 0.385. The temperature necessary to condense the methanol is 17.3°F, and the required removal efficiency is 90%. The coolant is ethylene glycol. The approach is 15°F, and the temperature rise of the coolant is 25°F. The heat capacity of the gaseous stream is 6.95 Btu/lbmol-°F. For methanol at 17.3°F, the latent heat is 17,230.6 Btu/lbmol. The emission stream is free of moisture and the ideal gas law applies. The total heat transferred from the air/methanol mixture to the ethylene glycol in the process is most nearly

(A) 14,400 Btu/hr
(B) 178,000 Btu/hr
(C) 183,000 Btu/hr
(D) 197,000 Btu/hr

Hint: Calculate the molar rate of methanol condensed. Perform an energy balance around the system to calculate the heat transfer.

PROBLEM 74

A distillation column separates a mixture consisting of 8% (by mole) acetone, 61.8% (by mole) acetic acid, and 30.2% (by mole) acetic anhydride. The column is designed to yield 825 kmol/h of a distillate stream containing 10% (by mole) acetone and 76% (by mole) acetic acid. The bottoms stream contains no acetone, and the column is operated so that 55% (by mole) of the overhead is returned as reflux. The mole fraction of acetic acid in the bottoms stream is most nearly

(A) 0.0019
(B) 0.050
(C) 0.57
(D) 0.60

Hint: Perform a mole balance of acetone around the system, an overall mole balance around the distillation column, and a mole balance of acetic acid around the splitter.

PROBLEM 75

A forced-draft cooling tower cools 16,000 gal/min of water from a temperature of 104.0°F to 89.0°F. The outlet air at these conditions is 100.0°F dry-bulb and 96.0°F wet-bulb. The air enters the tower at a flow rate of 80,800 lbm/min. The ambient wet-bulb temperature is 80°F, and the site altitude is sea level. The air operating line can be represented by the equation

$$h_a = -103.0606 \frac{\text{Btu}}{\text{lbm}} + \left(1.64863 \frac{\text{Btu}}{\text{lbm-°F}}\right) T$$
$$+ \left(2.96296 \times 10^{-6} \frac{\text{Btu}}{\text{lbm-°F}}\right) T^2$$

The water operating line can be represented by the equation

$$h_w = -77.9921 \frac{\text{Btu}}{\text{lbm}} + \left(1.4877 \frac{\text{Btu}}{\text{lbm-°F}}\right) T_w$$

The tower characteristic is most nearly

(A) 0.85 lbm-°F/Btu
(B) 1.1 lbm-°F/Btu
(C) 1.6 lbm-°F/Btu
(D) 6.3 lbm-°F/Btu

Hint: Calculate the enthalpy of the air-water-vapor mixture at four different wet-bulb temperatures and at four different bulk water temperatures. Use the Merkel equation to calculate the tower characteristic.

KINETICS

PROBLEM 76

For the hypothetical reaction $2F + M \rightarrow R$, the initial reaction rate is measured at several initial concentrations in a batch reactor.

experiment	C_F (mol/L)	C_M (mol/L)	initial rate of formation of R, r_R (mol/L·s)	temperature (°C)
1	0.25	0.25	1.5×10^{-3}	25
2	0.25	0.25	2.7×10^{-3}	30
3	0.10	0.10	4.0×10^{-4}	30
4	0.20	0.10	1.6×10^{-3}	30
5	0.30	0.10	3.6×10^{-3}	30
6	0.20	0.20	1.6×10^{-3}	30
7	0.30	0.30	3.6×10^{-3}	30

The activation energy for this reaction is most nearly

(A) 88 kJ/mol
(B) 11 000 kJ/mol
(C) 21 000 kJ/mol
(D) 110 000 kJ/mol

Hint: Calculate the ratio of rate constants at the two given temperatures. Apply the Arrhenius equation to calculate the frequency factor.

PROBLEM 77

The dimer alpha-methyl styrene ($C_{18}H_{20}$) decomposes in a reversible liquid-phase reaction to produce the monomers 2,4-diphenyl-4-methyl-1-pentene (C_9H_{10}) and 2,4-diphenyl-4-methyl-2-pentene (C_9H_{10}). The decomposition reaction is carried out in a constant-volume batch reactor at a constant temperature of 340K and pressure of 2 atm. The concentration equilibrium constant for this decomposition at 340K is 10 mol/L. The feed consists of $C_{18}H_{20}$ at a concentration of 7.174 mol/L at 340K and 2 atm. The equilibrium conversion of $C_{18}H_{20}$ is most nearly

(A) 0.0
(B) 0.30
(C) 0.92
(D) 0.95

Hint: Construct the stoichiometric table. Replace the concentrations in the equilibrium constant expression with concentrations from the stoichiometric table.

PROBLEM 78

For the first-order reaction $M \rightarrow R$, in a tubular reactor, the conversion of the species M is 85%. In the reactor, the volumetric flow rate is constant. At the design conditions, the flow rate is 12 L/min and the specific reaction rate is 0.258 min^{-1} at 25°C. In the reactor, the entering molar flow rate is 1.5 mol/L. At 25°C, the reactor volume is most nearly

(A) 0.04 L
(B) 7 L
(C) 90 L
(D) 300 L

Hint: Obtain an equation relating the reactor volume to the entering and exiting concentrations of M.

PROBLEM 79

For the elementary, irreversible, first-order reaction M→R, the conversion of species M entering an isothermal, continuous-flow tubular reactor into product R is 90%. In the reactor, the flow rate is constant at 158.5 gal/hr. At the design conditions, the specific reaction rate is 0.006 min^{-1} at 77°F. The feed to the reactor consists of pure M at a concentration of 3.785 lbmol/gal. At 77°F, the reactor volume is most nearly

(A) 0.005 gal
(B) 1000 gal
(C) 4000 gal
(D) 60,000 gal

Hint: Obtain an equation relating the reactor volume to the entering and exiting concentration of M.

PROBLEM 80

Ethanol (C_2H_5OH) at 675°C and 1 bar is fed to an adiabatic reactor where 35% of it is dehydrogenated to ethyl acetate ($C_4H_8O_2$) according to the irreversible reaction $2C_2H_5OH(g) \rightarrow C_4H_8O_2(g) + 2H_2(g)$. The heat capacities are constant over the range of temperature of the reaction. The heat capacities and standard heats of formation for ethanol, ethyl acetate, and hydrogen at 25°C are given in the following table.

component	heat capacity, $\overline{C}_p$ (cal/mol·°C)	standard heat of formation at 25°C, ΔH_f^0 (kcal/mol)
$C_2H_5OH(g)$	26.7	−52.23
$C_4H_8O_2(g)$	9.6	−102.02
$H_2(g)$	7.0	0

The temperature of the gases leaving the reactor is most nearly

(A) 5.10°C
(B) 810°C
(C) 830°C
(D) 850°C

Hint: The material and energy balances allow calculation of the outlet flows and of the temperature of the gases leaving the reactor.

PROBLEM 81

The decomposition of an organic compound is carried out in a longitudinal fixed-bed reactor operating isothermally at 720°C and 1.0 atm. At these conditions, the reaction constant for the decomposition is 0.0182 mol/L·s·atm$^{1.5}$ and the reaction order is 1.5. Assume the ideal gas law applies and assume constant volume. The concentration of the organic compound 3 s after entering the reactor is most nearly

(A) 1.2×10^{-3} mol/L
(B) 7.9×10^{-3} mol/L
(C) 9.9×10^{-3} mol/L
(D) 3.4×10^{-2} mol/L

Hint: Determine the reaction-rate equation.

PROBLEM 82

The reaction $M \rightarrow R + S$ is a vapor-phase reaction taking place on a solid catalyst. This reaction is carried out in a longitudinal fixed-bed reactor operating isothermally at 572°C and 0.2 atm. The feed is 20 kmol/h of pure M at 572°C. A conversion of 0.2 kmol of M/kmol of component M fed is achieved. Experimental data for a particular solid catalyst indicate that the reaction rate can be represented over a large temperature range by the rate equation

$$r_M = \frac{\left(0.0697 \, \frac{\text{kmol M converted}}{\text{kg of catalyst·atm·h}}\right) \times \left(p_M - \frac{p_R p_S}{0.203 \text{ atm}}\right)}{\left(1 + (0.445 \text{ atm}^{-1})p_M + (1.526 \text{ atm}^{-1})p_{RS}\right)^2}$$

variable	description	units
p_M	pressure of M	atm
p_R	pressure of R	atm
p_S	pressure of S	atm
p_{RS}	$(p_R + p_S)/2$	atm

The amount of catalyst required to achieve this conversion is most nearly

(A) 0.038 kg
(B) 30 kg
(C) 390 kg
(D) 430 kg

Hint: Determine the pressures and place their values in the given rate reaction. Replace the concentration in the rate reaction with the conversion, and integrate.

PROBLEM 83

The reversible gas-phase formation of ammonia (NH_3) from nitrogen (N_2) and hydrogen (H_2), $N_2(g) + 3H_2(g) \leftrightarrow 2NH_3(g)$, is carried out in a constant-volume batch reactor at constant temperature and pressure. The feed consists of 10 mol/L of N_2 and 30 mol/L of H_2 at 673K and 200 atm. The equilibrium concentration of N_2 at 673K is 0.5 mol/L. The equilibrium concentration constant for the formation of NH_3 is most nearly

(A) 0.0050 L²/mol²
(B) 11 L²/mol²
(C) 210 L²/mol²
(D) 480 L²/mol²

Hint: Replace the equilibrium concentrations of each of the species in the expression for the equilibrium constant.

PROBLEM 84

The irreversible liquid-phase reaction $P + Q \rightarrow R + S$ is carried out in a constant-volume batch reactor at constant temperature and pressure. The feed consists of 0.368 mol/L of P and 0.785 mol/L of Q. The conversion data are

time (min)	C_R (mol/L)
0	0
4	0.06
10	0.135
16	0.188
20	0.218
26	0.248

The rate law, in terms of the concentrations of P and Q, that best describes the data is most nearly

(A) $-r_P = (0.03 \text{ min}^{-1}) C_P C_Q$
(B) $-r_P = \left(0.07 \dfrac{\text{L}}{\text{mol·min}}\right) C_P C_Q$
(C) $-r_P = (0.03 \text{ min}^{-1}) C_P^2$
(D) $-r_P = \left(0.07 \dfrac{\text{L}}{\text{mol·min}}\right) C_P^2$

Hint: Determine the order of the reaction and the rate constant.

PROBLEM 85

The irreversible, first-order, liquid-phase reaction M→R is to be carried out in a series of two constant-volume continuous-stirred tank reactors (CSTRs). The two reactors are to be operated isothermally. The molar flow rate of M is 109 mol/min and the desired overall production of R is 87 mol/min. The initial concentration of M is 1 mol/L and the specific reaction rate constant is 0.41 min^{-1}. The minimum total volume of the reactor system is most nearly

(A) 320 L
(B) 520 L
(C) 650 L
(D) 840 L

Hint: Determine the conversion in the first reactor. With the outlet conversion from the second reactor, calculate the minimum volume.

PLANT DESIGN AND OPERATION

PROBLEM 86

The thermal cracking of acetone to produce 1.1×10^8 lbm/yr of acetic anhydride is to be carried out in a plug flow reactor at 1382°F. The reaction takes place in long coils placed in a furnace. These coils must be made of a material with good thermal and mechanical properties. The best material(s) for construction the coils is/are

I. copper
II. silica
III. nickel-free iron
IV. a high-chromium-iron alloy

(A) II only
(B) IV only
(C) I or II
(D) II or III

Hint: Consider the structural properties of the materials.

PROBLEM 87

A reagent, R, with a density of 0.0453 lbm/ft³ is fed into a plug flow reactor at a rate of 26.54 lbm/hr to generate a mixture of gaseous products. The desired conversion of R is 0.12. From experimental data, the ratio, z, of the volume of the reactor, V, in cubic feet, to the mass flow rate of the feed, $\dot{m}_F$, in pounds mass per hour is correlated to the conversion of R. The conversion of R is $X = 0.1916 \ln z - 0.0865$. The mean residence time is most nearly

(A) 0.01 hr
(B) 0.05 hr
(C) 0.1 hr
(D) 8 hr

Hint: Calculate the volume of the reactor and the volumetric feed rate.

PROBLEM 88

A phase diagram of the metallic elements M and N forming two solid phases, the α phase and the β phase, is shown on the following page. An alloy with 0.2% (by mole) N is cooled from 950°C to 800°C under equilibrium conditions.

At 800°C, the mole fraction of the solid phase is most nearly

(A) 0.1
(B) 0.3
(C) 0.4
(D) 0.6

Hint: Draw an isotherm at 800°C and use the lever rule on the segment determined by the equilibrium curves and the isotherm.

PROBLEM 89

500 lbm of a solution of a solid solute A in B (raffinate solvent) containing 65% (by weight) A is to be extracted two times using C as the solvent. The extraction is isothermal at 25°C. 185 lbm of solvent C are to be used in each extraction. Use the equilibrium diagram provided on the following page.

Illustration for Problem 88

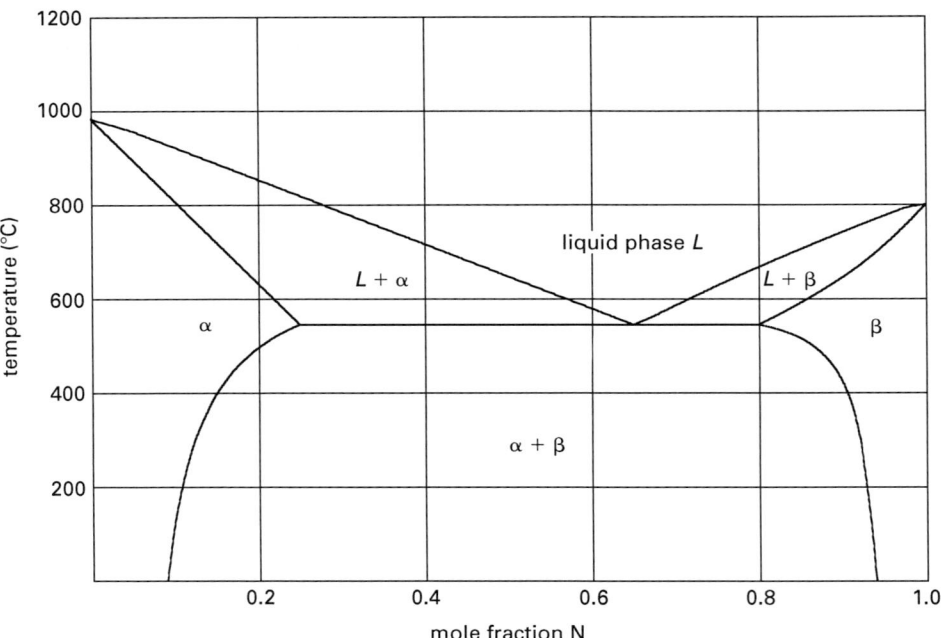

Illustration for Problem 89

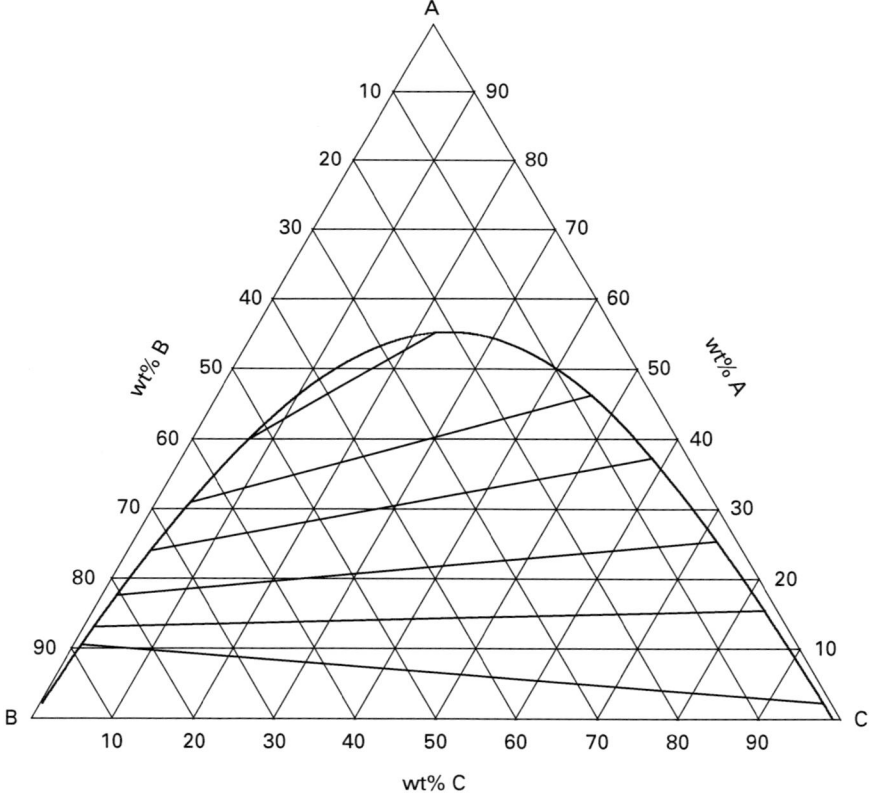

The mass fraction of A extracted in the second stage is most nearly

(A) 0.14
(B) 0.18
(C) 0.21
(D) 0.52

Hint: Use the diagram provided to calculate the masses of the raffinate and extract in the first and second stages.

PROBLEM 90

A test burn waste mixture consisting of chlorobenzene (C_6H_5Cl), toluene (C_7H_8), and xylene (C_8H_{10}) is incinerated at 1000°C. The stack gas flow rate is 375.24 dscm/min (dry standard cubic meters per minute). To comply with Environmental Protection Agency (EPA) regulations, the HCl emission may not exceed the greater of 1.8 kg/h or 1% (by weight) of the HCl present prior to the installation of the air pollution control equipment. The destruction and removal efficiency (DRE) for C_6H_5Cl, C_7H_8, and C_8H_{10} must be 99.99% (by weight) to comply with federal regulations. All the chlorine in the feed is converted to HCl. The waste feed rate and the stack discharge are shown in the following table.

compound	molecular weight (g/mol)	$\dot{m}_{in}$ (kg/h)	$\dot{m}_{out}$ (kg/h)
C_6H_5Cl	112.5	154	0.012
C_7H_8	92.0	431	0.036
C_8H_{10}	106	434	0.070
HCl	36.5	–	1.1

Which of the following emissions is/are in compliance with EPA regulations?

I. C_6H_5Cl
II. C_7H_8
III. C_8H_{10}
IV. HCl

(A) III only
(B) I and II
(C) II and III
(D) I, II, and IV

Hint: Calculate DRE for each component—C_6H_5Cl, C_7H_8, and C_8H_{10}. Check compliance for HCl emission.

PROBLEM 91

A reactor with a capacity of 400 ft³, initially filled with ambient air, is to handle powdered solids. To reduce the possibility of explosion due to rapid oxidation of the powdered solids, the concentration of oxygen in the reactor is to be controlled using nitrogen at 90°F. The safe final concentration of oxygen in the reactor is 4% (by volume). Assume the ideal gas law applies. The mass of nitrogen required, using sweep through purging, is most nearly

(A) 30 lbm
(B) 50 lbm
(C) 300 lbm
(D) 700 lbm

Hint: Consider the nitrogen being added to the reactor at one end and withdrawn at the other end.

PROBLEM 92

An 8 ft diameter conical-bottom tank is filled with a liquid. The volume of the conical portion of the tank is 157.2 ft³, and the height of the conical portion of the tank is 2.31 ft. The discharge piping is 1.5 in schedule-80 steel pipe having an internal diameter of 0.125 ft. The discharge piping consists of 200 ft of pipe plus two long-radius elbows with an equivalent length of 14 ft and one angle valve with an equivalent length of 55 ft. The discharge-piping outlet is 12 ft below the bottom of the tank. The initial liquid level in the tank above the tank bottom is 15 ft. The Darcy friction factor is 0.03160. The drainage times (in minutes) in the cylindrical, t_{cyl}, and conical, t_{con}, sections of the tank are given by the following equations.

$$t_{cyl, minutes} = 24.058\sqrt{1+K}$$
$$t_{con, minutes} = 1.768\sqrt{1+K}$$

K is the net resistance coefficient. The average fluid velocity at the drain-pipe outlet is

(A) 3.9 ft/sec
(B) 4.9 ft/sec
(C) 6.0 ft/sec
(D) 300 ft/sec

Hint: Calculate the draining time for the cylindrical and conical portions of the tank. Calculate the total time and the volume drained out in that time.

PROBLEM 93

A saturated liquid feed containing 40% (by mole) trichloroethene and 60% (by mole) tetrachlorethene is fractionated at 760 mm Hg to produce 96% (by mole) trichloroethene and a residue containing 98% (by mole) tetrachloroethene. The reflux ratio is 2.1 and the relative volatility is 2.65. Assume constant molal overflow. Counting from the top of the column, the feed plate that will achieve this separation is

(A) 5
(B) 6
(C) 7
(D) 8

Hint: Perform a total mole balance around the column together with a trichloroethene mole balance to determine the rectifying and feed lines.

PROBLEM 94

An equalization tank containing water with dissolved ions is to be connected to a process tank. A pump is required because the process tank liquid elevation is 40 m above the equalization tank level. The piping system requires one gate valve, two swing check valves, four standard 90° elbows, and 70 m of piping. The process tank temperature is 25°C and its flow rate is 0.05 m³/s. The piping material has a roughness of 1.4×10^{-6} m. The water has a normal service velocity of 2.1 m/s and a kinematic viscosity of 9.01×10^{-7} m²/s. For the piping system components, the individual resistance coefficients are as follows.

minor loss	resistance coefficient
entry	0.5
gate valve	0.2
swing check valve	2.5
elbow	0.35
exit	1.0

The required pump head is most nearly

(A) 43 m
(B) 45 m
(C) 46 m
(D) 98 m

Hint: First calculate the pipe size and then calculate the pump head.

PROBLEM 95

A waste gas entering an absorber contains hydrochloric acid. The solvent is a caustic solution with the same physical properties as water. The conditions of the absorber are

condition	symbol	liquid
molecular weight	MW_L	18 lbm/lbmol
	MW_G	29 lbm/lbmol
density	ρ_L	62.4 lbm/ft³
	ρ_G	0.0709 lbm/ft³
viscosity	μ_L	2.16 lbm/ft-hr
	μ_G	–
flow rate	$L_{mol,i}$	7275 lbmol/hr
	$G_{mol,i}$	3194 lbmol/hr

The absorber is randomly packed with 2 in ceramic Raschig rings having a packing factor of 65. Eckert's flooding curve expresses the abscissa as

$$X = \frac{L_{mol,i} MW_L}{G_{mol,i} MW_G} \sqrt{\frac{\rho_G}{\rho_L - \rho_G}}$$

Unit Operations of Chemical Engineering expresses the ordinate as

$$Y = \frac{G_{sfr,i}^2 F_p \mu_L^{0.1}}{(\rho_L - \rho_G) \rho_G g_c}$$

At the conditions of the absorber, the ordinate may also be calculated as

$$Y = e^{\left(\begin{array}{c}-4.026 - 0.9895 \ln X - 0.0829 (\ln X)^2 \\ +0.0324 (\ln X)^3 + 0.0053 (\ln X)^4\end{array}\right)}$$

The column diameter is most nearly

(A) 8.0 ft
(B) 8.2 ft
(C) 20 ft
(D) 490 ft

Hint: Calculate the abscissa and ordinate of the flooding curve. Use the Eckert modification to calculate the superficial gas flow rate entering the absorber.

PROBLEM 96

A distillate product containing 1218 mol/min acetone and 36 mol/min water is to be obtained from a feed material at its bubble point. The column is operated at 105 kPa at the top of the tower. The distillate is produced at 322K and the bottoms are produced at 382K. The reflux ratio is 1.10. The distillation column has 24 in tray spacing. The fraction of the area available for vapor flow is 0.60. The liquid phase has a density of 10^6 g/m³ and a molecular weight of 18 g/mol. The allowable percent of flooding is 80%. The surface tension is 70×10^{-2} N/m. Assume a pressure drop of 50 kPa and that the ideal gas law applies. For the 24 in tray spacing, the correlation between the ordinate of the flooding curve, C_{sb}, in meters per second, and the abscissa, a dimensionless flow parameter, is

$$C_{sb} = e^{-3.46 + \sqrt{1.69 - 2.03 \ln F_{lv} - 0.217(\ln F_{lv})^2}}$$

$$F_{lv} = \frac{\dot{m}_{L'}}{\dot{m}_{V'}} \sqrt{\frac{\rho_g}{\rho_L}}$$

The ordinate of the flooding curve is

$$C_{sb} = U_{nf} \left(\frac{20 \times 10^{-2} \frac{N}{m}}{\sigma}\right)^{0.2} \sqrt{\frac{\rho_g}{\rho_L - \rho_g}}$$

D
p_1 = 105 kPa
T_D = 322K
D_A = 1218 mol/min acetone
D_w = 36 mol/min water

feed

B
T_B = 382K
B_A = 282 mol/min acetone
B_w = 20064 mol/min water

The column diameter for the section below the feed tray that will achieve the desired separation is most nearly

(A) 0.12 m
(B) 0.38 m
(C) 0.49 m
(D) 3.8 m

Hint: Calculate the liquid and vapor flow rates in the bottom section of the column. Use the flooding curve to calculate the velocity of the fluids in the bottom section.

PROBLEM 97

The hydrolysis of ethylene (C_2H_4) is to be considered an irreversible reaction at constant volume for design purposes. The reaction is first order with respect to C_2H_4 and zero order with respect to water. From experimental data, the correlation of the reaction rate constant and the absolute temperature is

$$k = (-0.0145 \text{K}^{-1} \text{min}^{-1})T + 9.682 \text{min}^{-1}$$

At the same time, the correlation between the conversion of C_2H_4 and the absolute temperature is

$$X = (0.0012 \text{K}^{-1})T + 0.5$$

The reaction is to be carried out at 30°C in two continuously stirred tank reactors (CSTRs) in series, both having the same volume. The throughput is 374 gal/min. The total reactor volume required is most nearly

(A) 70 gal
(B) 120 gal
(C) 450 gal
(D) 1200 gal

Hint: Consider the conversion to calculate the space time in a series of two CSTRs and a first-order reaction.

PROBLEM 98

A saturated liquid feed containing 45% (by mole) of a chlorinated hydrocarbon (C1) and 55% (by mole) of a second chlorinated hydrocarbon (C2) is fractionated at 760 mm Hg to produce 97% (by mole) C1 and a residue containing 97% (by mole) C2. The reflux ratio is 2.81, and the relative volatility is 3.2. The number of equilibrium plates needed to achieve this separation is

(A) 4
(B) 7
(C) 8
(D) 9

Hint: Perform a total mole balance around the column together with a C1 mole balance to determine the rectifying, stripping, and q lines.

PROBLEM 99

A wastewater stream flows at 132 m³/d into a 12.1 m³ aeration basin. From the aeration basin the stream flows into a clarifier. The soluble chemical oxygen demand (COD) of the feed is 0.11 kg/m³. There are 0.09 kg/m³ of biological suspended solids in the influent stream, reported as COD. The clarifier overflow soluble COD is 0.032 kg/m³. The clarifier overflow contains no biological suspended solids. The concentration of the biological suspended solids in the underflow of the clarifier, reported as COD, is 9 kg/m³. The fraction of biological solids converted to biomass, calculated on a COD basis, is 0.53. The mean biological solids retention time is 2.8 d. The specific biomass loss rate for biomass fed equals the specific biomass loss rate for solids formed in the basin, which is 0.034/d. Assume no significant evaporation in the clarifier and steady-state treatment.

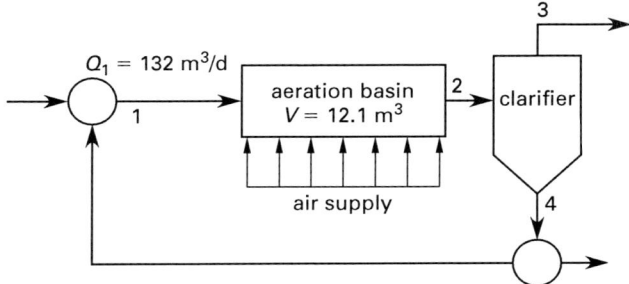

The concentration of oxygen required is most nearly

(A) 4.3 kg/d
(B) 6.3 kg/d
(C) 16 kg/d
(D) 28 kg/d

Hint: Perform a mass balance around the wastewater treatment and around the clarifier.

PROBLEM 100

A wastewater contains 125 mg/L CO_3^{2-} and 82 mg/L HCO_3^- at a pH of 9. At 25°C, the ion product of water is 10^{-14} mol²/L².

species	molecular weight (g/mol)
Ca	40
C	12
H	1
O	16

At 25°C, the alkalinity as calcium carbonate ($CaCO_3$) is most nearly.

(A) 276 mg/L
(B) 343 mg/L
(C) 350 mg/L
(D) 484 mg/L

Hint: Convert the concentrations of CO_3^{2-}, HCO_3^-, OH^-, and H^+ to concentrations as $CaCO_3$.

Solutions

MASS/ENERGY BALANCES AND THERMODYNAMICS

SOLUTION 1

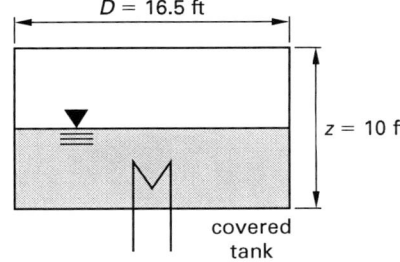

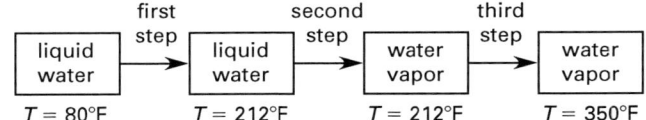

The volume of the tank is

$$V = \left(\frac{\pi D^2}{4}\right) z = \left(\frac{\pi (16.5 \text{ ft})^2}{4}\right)(10 \text{ ft})$$
$$= 2138 \text{ ft}^3$$

Because the tank is half-filled, the volume occupied by the water in the tank is

$$V_{\text{H}_2\text{O}} = \frac{V}{2} = \frac{2138 \text{ ft}^3}{2}$$
$$= 1069 \text{ ft}^3$$

The density of water, ρ, is given as 62.21 lbm/ft^3. The mass of water in the tank is

$$m = \rho V_{\text{H}_2\text{O}} = \left(62.21 \, \frac{\text{lbm}}{\text{ft}^3}\right)(1069 \text{ ft}^3)$$
$$= 66{,}500 \text{ lbm}$$

Because enthalpy is a function of state, the entire process may be considered as if it happened in three steps. In the first step, the liquid water goes from a temperature of 80°F to 212°F and there are no phase transitions. The enthalpy change of the liquid water in the first step is calculated as the enthalpy change of liquid water going from 80°F to 212°F. The second step involves an isothermal change of phase from liquid at 212°F to vapor at 212°F. The enthalpy change in the second step is identical to the heat of vaporization. In the last step, the water vapor at 212°F goes to water vapor at 350°F.

For the first step, the enthalpy change per pound of the liquid water from 80°F to 212°F is

$$\Delta h_{80 \to 212} = c_{p,l}(T_{212} - T_{80})$$
$$= \left(1 \, \frac{\text{Btu}}{\text{lbm-°F}}\right)(212°\text{F} - 80°\text{F})$$
$$= 132 \text{ Btu/lbm}$$

In the second step, the enthalpy change equals the heat of vaporization of the water at 212°F because this step involves a change of phase at constant temperature.

The heat of vaporization of water at 212°F, ΔH_{212}, is given as 970.3 Btu/lbm. In the third step, the enthalpy change per pound of the water vapor from 212°F to water vapor at 350°F is

$$\Delta h_{212 \to 350} = c_{p,v}(T_{350} - T_{212})$$
$$= \left(0.48 \, \frac{\text{Btu}}{\text{lbm-°F}}\right)(350°\text{F} - 212°\text{F})$$
$$= 66.24 \text{ Btu/lbm}$$

The enthalpy changes for the individual steps sum to give the total enthalpy change. The total enthalpy change per pound of water from 80°F to 350°F is

$$\Delta h_{80 \to 350} = \Delta h_{80 \to 212} + \Delta h_{212} + \Delta h_{212 \to 350}$$
$$= 132 \, \frac{\text{Btu}}{\text{lbm}} + 970.3 \, \frac{\text{Btu}}{\text{lbm}} + 66.24 \, \frac{\text{Btu}}{\text{lbm}}$$
$$= 1168.54 \text{ Btu/lbm}$$

The total enthalpy change is

$$(\Delta h_{80 \to 350})\, m = \left(1168.54 \, \frac{\text{Btu}}{\text{lbm}}\right)(66{,}500 \text{ lbm})$$
$$= 78 \times 10^6 \text{ Btu}$$

The answer is (C).

Why Other Options Are Wrong

(A) When calculating the total enthalpy change, this incorrect answer omits the enthalpy change when the liquid water changes state.

(B) When calculating the total enthalpy change, this incorrect answer omits the enthalpy change required to increase the temperature of the water from 80°F to 212°F.

(D) In this incorrect answer, the tank volume is used instead of the water volume when calculating the mass of liquid water.

SOLUTION 2

The vapor pressure of vinyl chloride is given as

$$p_{\text{vinyl}} = 5.77 \text{ lbf/in}^2$$

The total pressure, p, is given as 14.3 lbf/in². The vapor mole fraction of vinyl chloride is

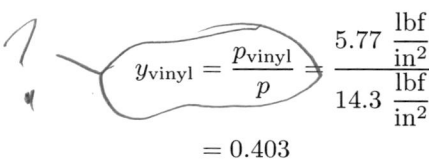

$$y_{\text{vinyl}} = \frac{p_{\text{vinyl}}}{p} = \frac{5.77 \frac{\text{lbf}}{\text{in}^2}}{14.3 \frac{\text{lbf}}{\text{in}^2}}$$
$$= 0.403$$

The molecular weight of vinyl chloride, MW_{vinyl}, is given as 62.5 lbm/lbmol. The molecular weight of dry air, MW_{air}, is given as 28.96 lbm/lbmol. The mass fraction of vinyl chloride is

$$w_{\text{vinyl}} = \frac{y_{\text{vinyl}}(\text{MW}_{\text{vinyl}})}{y_{\text{vinyl}}(\text{MW}_{\text{vinyl}}) + (1 - y_{\text{vinyl}})(\text{MW}_{\text{air}})}$$

$$= \frac{(0.403)\left(62.5 \frac{\text{lbm}}{\text{lbmol}}\right)}{(0.403)\left(62.5 \frac{\text{lbm}}{\text{lbmol}}\right) + (1 - 0.403)\left(28.96 \frac{\text{lbm}}{\text{lbmol}}\right)}$$

$$= 0.593$$

Because the saturated mixture contains only two components, air and vinyl chloride, the mass ratio of vinyl chloride to dry air is

$$\text{ratio} = \frac{w_{\text{vinyl}}}{1 - w_{\text{vinyl}}} = \frac{0.593}{1 - 0.593}$$
$$= 1.46$$

The answer is (D).

Why Other Options Are Wrong

(A) This incorrect answer uses the mole fraction (0.403) instead of the mass fraction (0.593) in computing the mass ratio of vinyl chloride to dry air.

(B) This incorrect answer uses the molecular weight of air instead of the molecular weight of vinyl chloride in calculating the mass fraction of vinyl chloride.

(C) This incorrect answer uses the standard pressure instead of the total pressure when calculating the mass fraction of vinyl chloride.

SOLUTION 3

The vapor pressure of vinyl chloride, p_{vinyl}, is given as 2580 mm Hg. The total pressure of the mixture, p, is given as 3739 mm Hg. The molecular weight of vinyl chloride, MW_{vinyl}, is given as 62.50 g/mol. The absolute temperature of the mixture is

$$T = 20°\text{C} + 273.15°$$
$$= 293.15 \text{K}$$

The universal gas constant is

$$R = 62.361 \text{ mm Hg·L/mol·K}$$
$$(62\,361 \text{ mm Hg·cm}^3/\text{mol·K})$$

The density of vinyl chloride using the ideal gas law is

$$\rho_{\text{vinyl}} = \frac{p_{\text{vinyl}}(\text{MW}_{\text{vinyl}})}{RT}$$

$$= \frac{(2580 \text{ mm Hg})\left(62.50 \frac{\text{g}}{\text{mol}}\right)}{\left(62\,361 \frac{\text{mm Hg·cm}^3}{\text{mol·K}}\right)(293.15 \text{K})}$$

$$= 0.00882 \text{ g/cm}^3$$

In 100 cm³ of mixture, the mass of vinyl chloride is

$$m = \rho_{\text{vinyl}} V$$
$$= \left(0.00882 \frac{\text{g}}{\text{cm}^3}\right)(100 \text{ cm}^3)$$
$$= 0.882 \text{ g} \quad (0.88 \text{ g})$$

The answer is (C).

Why Other Options Are Wrong

(A) This incorrect answer is the mass of vinyl chloride in 1 cm³ of mixture instead of in 100 cm³ of mixture.

(B) For this incorrect answer, the density of vinyl chloride is calculated using the molecular weight of dry air instead of that of vinyl chloride.

(D) This incorrect answer calculates the density using the temperature in degrees Celsius instead of the absolute temperature.

SOLUTION 4

The mass flow rate of the product, $\dot{m}_{\text{out}}$, is given as 48,000 lbm/hr. The mass fraction of P in the product, $w_{\text{P,out}}$, is given as 0.04. The mass flow rate of P in the product is

$$\dot{m}_{\text{P,out}} = w_{\text{P,out}} \dot{m}_{\text{out}}$$
$$= (0.04)\left(48{,}000 \ \frac{\text{lbm}}{\text{hr}}\right)$$
$$= 1920 \ \text{lbm/hr}$$

The mass fraction of M in the product, $w_{\text{M,out}}$, is given as 0.22. The mass flow rate of M in the product is

$$\dot{m}_{\text{M,out}} = w_{\text{M,out}} \dot{m}_{\text{out}}$$
$$= (0.22)\left(48{,}000 \ \frac{\text{lbm}}{\text{hr}}\right)$$
$$= 10{,}560 \ \text{lbm/hr}$$

Because the only source of M is stream J1, the mass flow rate of M in J1 is the same as the mass flow rate of M in the product. The mass flow rate of M in stream J1 is

$$\dot{m}_{\text{M,J1}} = \dot{m}_{\text{M,out}}$$
$$= 10{,}560 \ \text{lbm/hr}$$

The mass flow rate of P in stream J2 is the same as the mass flow rate of P in the product because J2 is the sole source of component P. The mass flow rate of P in stream J2 is

$$m_{\text{P,J2}} = \dot{m}_{\text{P,out}}$$
$$= 1920 \ \text{lbm/hr}$$

The mass fraction of P in stream J2, $w_{\text{P,J2}}$, is given as 0.95. The mass flow rate of stream J2 equals the mass flow rate of P in J2 divided by the mass fraction of P in J2.

$$\dot{m}_{\text{J2}} = \frac{\dot{m}_{\text{P,J2}}}{w_{\text{P,J2}}} = \frac{1920 \ \frac{\text{lbm}}{\text{hr}}}{0.95}$$
$$= 2021 \ \text{lbm/hr}$$

Because stream J2 has only two components, P and water, the mass fraction of water in J2 is

$$w_{\text{W,J2}} = 1 - w_{\text{P,J2}} = 1 - 0.95$$
$$= 0.05$$

The mass flow rate of water in stream J2 is

$$\dot{m}_{\text{W,J2}} = w_{\text{W,J2}} \dot{m}_{\text{M,J2}} = (0.05)\left(2021 \ \frac{\text{lbm}}{\text{hr}}\right)$$
$$= 101 \ \text{lbm/hr}$$

The mass fraction of M in stream J1, $w_{\text{M,J1}}$, is given as 0.99. The mass flow rate of stream J1 is

$$\dot{m}_{\text{J1}} = \frac{\dot{m}_{\text{M,J1}}}{w_{\text{M,J1}}} = \frac{10{,}560 \ \frac{\text{lbm}}{\text{hr}}}{0.99}$$
$$= 10{,}667 \ \text{lbm/hr}$$

Because stream J1 has only two components, M and water, the mass fraction of water in J1 is

$$w_{\text{W,J1}} = 1 - w_{\text{M,J1}} = 1 - 0.99$$
$$= 0.01$$

The mass flow rate of water in stream J1 is

$$\dot{m}_{\text{W,J1}} = w_{\text{W,J1}} \dot{m}_{\text{J1}} = (0.01)\left(10{,}667 \ \frac{\text{lbm}}{\text{hr}}\right)$$
$$= 107 \ \text{lbm/hr}$$

A mass balance around mixer 1 gives the mass flow rate of water in stream S1. The mass flow rate of water in S1 is

$$\dot{m}_{\text{W,S1}} = \dot{m}_{\text{W,J1}} + \dot{m}_{\text{W,J2}} = 107 \ \frac{\text{lbm}}{\text{hr}} + 101 \ \frac{\text{lbm}}{\text{hr}}$$
$$= 208 \ \text{lbm/hr}$$

The mass flow rate of stream S1 is the sum of the mass flow rates of M, P, and water. The mass flow rate of S1 is

$$\dot{m}_{\text{S1}} = \dot{m}_{\text{M,out}} + \dot{m}_{\text{P,out}} + \dot{m}_{\text{W,S1}}$$
$$= 10{,}560 \ \frac{\text{lbm}}{\text{hr}} + 1920 \ \frac{\text{lbm}}{\text{hr}} + 208 \ \frac{\text{lbm}}{\text{hr}}$$
$$= 12{,}688 \ \text{lbm/hr}$$

The percentage of P in stream S1 is the mass flow rate of P in S1 divided by the total mass flow rate of S1. The percentage of P in S1 is

$$\dot{m}_{\text{P,S1}} = \frac{\dot{m}_{\text{P,J2}}}{\dot{m}_{\text{S1}}} \times 100\% = \frac{1920 \ \frac{\text{lbm}}{\text{hr}}}{12{,}688 \ \frac{\text{lbm}}{\text{hr}}} \times 100\%$$
$$= 15.13\% \quad (15\%)$$

The answer is (B).

Why Other Options Are Wrong

(A) This incorrect answer is the mass fraction of water in stream S1.

(C) This incorrect answer is the percentage of water in the product.

(D) This incorrect answer is the mass fraction of M in stream S1.

SOLUTION 5

The mass flow rate of the product, $\dot{m}_{\text{out}}$, is given as 36,000 lbm/hr. The mass fraction of R in the product, $w_{\text{R,out}}$, is given as 0.24. The mass fraction of P in the product, $w_{\text{P,out}}$, is given as 0.040. The mass fraction of M in the product, $w_{\text{M,out}}$, is given as 0.20. The mass fraction of water in the product is

$$\begin{aligned} w_{\text{W,out}} &= 1 - w_{\text{M,out}} - w_{\text{P,out}} - w_{\text{R,out}} \\ &= 1 - 0.20 - 0.04 - 0.24 \\ &= 0.52 \end{aligned}$$

The mass flow rate of water in the product is

$$\begin{aligned} \dot{m}_{\text{W,out}} &= w_{\text{W,out}} \dot{m}_{\text{out}} \\ &= (0.52)\left(36{,}000 \ \frac{\text{lbm}}{\text{hr}}\right) \\ &= 18{,}720 \ \text{lbm/hr} \end{aligned}$$

The mass flow rate of P in the product is

$$\begin{aligned} \dot{m}_{\text{P,out}} &= w_{\text{P,out}} \dot{m}_{\text{out}} \\ &= (0.04)\left(36{,}000 \ \frac{\text{lbm}}{\text{hr}}\right) \\ &= 1440 \ \text{lbm/hr} \end{aligned}$$

The mass flow rate of M in the product is

$$\begin{aligned} \dot{m}_{\text{M,out}} &= w_{\text{M,out}} \dot{m}_{\text{out}} \\ &= (0.20)\left(36{,}000 \ \frac{\text{lbm}}{\text{hr}}\right) \\ &= 7200 \ \text{lbm/hr} \end{aligned}$$

The mass flow rate of R in the product is

$$\begin{aligned} \dot{m}_{\text{R,out}} &= w_{\text{R,out}} \dot{m}_{\text{out}} \\ &= (0.24)\left(36{,}000 \ \frac{\text{lbm}}{\text{hr}}\right) \\ &= 8640 \ \text{lbm/hr} \end{aligned}$$

The mass flow rate of R in stream J3 is the same as the mass flow rate of R in the product because J3 is the sole source of R in the product. The mass flow rate of R in J3 is

$$\begin{aligned} \dot{m}_{\text{R,J3}} &= \dot{m}_{\text{R,out}} \\ &= 8640 \ \text{lbm/hr} \end{aligned}$$

The mass fraction of R in J3, $w_{\text{R,J3}}$, is given as 0.48. Because stream J3 has only two components, R and water, the mass fraction of water in J3 is

$$\begin{aligned} w_{\text{W,J3}} &= 1 - w_{\text{R,J3}} = 1 - 0.48 \\ &= 0.52 \end{aligned}$$

The mass flow rate of water in J3 is

$$\begin{aligned} \dot{m}_{\text{W,J3}} &= \dot{m}_{\text{J3}} w_{\text{W,J3}} \\ &= \left(\frac{\dot{m}_{\text{R,J3}}}{w_{\text{R,J3}}}\right) w_{\text{W,J3}} \\ &= \left(\frac{8640 \ \frac{\text{lbm}}{\text{hr}}}{0.48}\right)(0.52) \\ &= 9360 \ \text{lbm/hr} \end{aligned}$$

The mass flow rate of M in stream J1 is the same as the mass flow rate of M in the product because J1 is the only source of M in the product. The mass flow rate of M in J1 is

$$\begin{aligned} \dot{m}_{\text{M,J1}} &= \dot{m}_{\text{M,out}} \\ &= 7200 \ \text{lbm/hr} \end{aligned}$$

The mass fraction of M in J1, $w_{\text{M,J1}}$, is given as 0.95. The mass flow rate of J1 is

$$\dot{m}_{\text{J1}} = \frac{\dot{m}_{\text{M,J1}}}{w_{\text{M,J1}}} = \frac{7200 \ \frac{\text{lbm}}{\text{hr}}}{0.95}$$
$$= 7579 \ \text{lbm/hr}$$

Because there are only two components in stream J1, M and water, the mass fraction of water in J1 is

$$\begin{aligned} w_{\text{W,J1}} &= 1 - w_{\text{M,J1}} = 1 - 0.95 \\ &= 0.05 \end{aligned}$$

The mass flow rate of water in J1 is

$$\begin{aligned} \dot{m}_{\text{W,J1}} &= w_{\text{W,J1}} \dot{m}_{\text{J1}} \\ &= (0.05)\left(7579 \ \frac{\text{lbm}}{\text{hr}}\right) \\ &= 379 \ \text{lbm/hr} \end{aligned}$$

The mass flow rate of P in stream J2 is the same as the mass flow rate of P in the product because J2 is the only source of component P in the product. The mass flow rate of P in J2 is

$$\dot{m}_{P,J2} = \dot{m}_{P,\text{out}}$$
$$= 1440 \text{ lbm/hr}$$

The mass fraction of P in J2, $w_{P,J2}$, is given as 0.95. The mass flow rate of J2 is

$$\dot{m}_{J2} = \frac{\dot{m}_{P,J2}}{w_{P,J2}} = \frac{1440 \frac{\text{lbm}}{\text{hr}}}{0.95}$$
$$= 1516 \text{ lbm/hr}$$

Because there are only two components in stream J2, component P and water, the mass fraction of water in J2 is

$$w_{W,J2} = 1 - w_{P,J2} = 1 - 0.95$$
$$= 0.05$$

The mass flow rate of water in J2 is

$$\dot{m}_{W,J2} = w_{W,J2}\dot{m}_{J2}$$
$$= (0.05)\left(1516 \frac{\text{lbm}}{\text{hr}}\right)$$
$$= 75.8 \text{ lbm/hr}$$

A water mass balance around mixer 1 gives the mass flow rate of water in stream S1. The mass flow rate of water in S1 is

$$\dot{m}_{W,S1} = \dot{m}_{W,J1} + \dot{m}_{W,J2}$$
$$= 379 \frac{\text{lbm}}{\text{hr}} + 75.8 \frac{\text{lbm}}{\text{hr}}$$
$$= 455 \text{ lbm/hr}$$

A water mass balance around mixer 2 gives the mass flow rate of water in stream J4. The mass flow rate of water in J4 is

$$\dot{m}_{W,J4} = \dot{m}_{W,\text{out}} - \dot{m}_{W,S1} - \dot{m}_{W,J3}$$
$$= 18{,}720 \frac{\text{lbm}}{\text{hr}} - 455 \frac{\text{lbm}}{\text{hr}} - 9360 \frac{\text{lbm}}{\text{hr}}$$
$$= 8905 \text{ lbm/hr} \quad (8900 \text{ lbm/hr})$$

The answer is (A).

Why Other Options Are Wrong

(B) When calculating the mass flow rate of water in stream J4, the mass flow rate of water in S1 is added instead of subtracted, resulting in this incorrect solution.

(C) When calculating the mass flow rate of water in stream J4, the mass flow rate of water in J3 is added instead of subtracted, resulting in this incorrect solution.

(D) When calculating the mass flow rate of water in stream J4, the mass flow rates of water in both J3 and S1 are added instead of subtracted, resulting in this incorrect solution.

SOLUTION 6

The molar flow rate of M in the feed, F_{M_F}, is given as 100 lbmol/hr. Mixer 1 receives the feed and the recycle streams. A mass balance of M around mixer 1 gives the molar flow rate of M entering the reactor. The molar flow rate of M entering the reactor is

$$F_{M_1} = F_{M_F} + F_{M_R}$$

Substituting the value of F_{M_F} in the preceding equation gives

$$F_{M_1} = 100 \frac{\text{lbmol}}{\text{hr}} + F_{M_R}$$

The mole fraction of M consumed in the reactor, x_{reacted}, is given as 0.98. The molar flow rate of M consumed in the reactor is given as

$$F_{M_{\text{consumed}}} = x_{\text{reacted}} F_{M_1}$$
$$= 0.98 F_{M_1}$$

By mass balance around the reactor, the molar flow rate of M leaving the reactor is the molar flow rate of M entering the reactor minus the molar flow rate of M that reacted. The molar flow rate of M leaving the reactor is

$$F_{M_S} = F_{M_1} - x_{\text{reacted}} F_{M_1}$$
$$= (1 - x_{\text{reacted}}) F_{M_1}$$
$$= (1 - 0.98) F_{M_1}$$
$$= (0.02)\left(100 \frac{\text{lbmol}}{\text{hr}} + F_{M_R}\right)$$
$$= 2 \frac{\text{lbmol}}{\text{hr}} + 0.02 F_{M_R}$$

The molar flow rate of M leaving the reactor equals the molar flow rate of M entering the separator because there are no materials added or transformed between the reactor and the separator.

The product stream contains 0.6% of total M entering the separator. The fraction of M in the product, x_{M_p}, is given as 0.006. The molar flow rate of M in the product is

$$F_{M_p} = x_{M_p} F_{M_S}$$
$$= 0.006 F_{M_S}$$

Replacing the expression for F_{M_S} previously obtained,

$$F_{M_P} = (0.006)(0.02)\left(100\ \frac{\text{lbmol}}{\text{hr}} + F_{M_R}\right)$$

$$= 0.012\ \frac{\text{lbmol}}{\text{hr}} + 0.00012 F_{M_R}$$

A mass balance of M around the separator gives the molar flow rate of M in the recycle stream. The molar flow rate of M in the recycle stream is

$$F_{M_R} = F_{M_S} - F_{M_P}$$

Substituting F_{M_S} and F_{M_P} in the preceding equation yields an expression where the only unknown is F_{M_R}.

$$F_{M_R} = 2\ \frac{\text{lbmol}}{\text{hr}} + 0.02 F_{M_R}$$
$$- \left(0.012\ \frac{\text{lbmol}}{\text{hr}} + 0.00012 F_{M_R}\right)$$

$$= \frac{2\ \frac{\text{lbmol}}{\text{hr}} - 0.012\ \frac{\text{lbmol}}{\text{hr}}}{1 - 0.02 + 0.00012}$$

$$= 2.03\ \text{lbmol/hr}$$

A mass balance of component N around mixer 1 gives the molar flow rate of N to the reactor. Because the feed does not contain N, the molar flow rate of N to the reactor equals the molar flow rate of N in the recycle stream. From the stoichiometry of the reaction, the moles of N formed in the reaction are twice the moles of M that reacted. The molar flow rate of N formed during the reaction is

$$F_{N_{\text{produced}}} = 2 F_{M_{\text{consumed}}}$$
$$= (2)(0.98)\left(100\ \frac{\text{lbmol}}{\text{hr}} + F_{M_R}\right)$$

Replacing the value of F_{M_R} in the preceding equation gives

$$F_{N_{\text{produced}}} = (2)(0.98)\left(100\ \frac{\text{lbmol}}{\text{hr}} + 2.03\ \frac{\text{lbmol}}{\text{hr}}\right)$$
$$= 199.98\ \text{lbmol/hr}$$

A mass balance of N around the reactor gives the molar flow rate of N leaving the reactor. The molar flow rate of N in the recycled stream is F_{N_R}. The molar flow rate of N leaving the reactor is the molar flow rate of N entering the reactor plus the molar flow rate of N formed in the reaction. The molar flow rate of N leaving the reactor is

$$F_{N_S} = F_{N_R} + F_{N_{\text{produced}}}$$
$$= F_{N_R} + 199.98\ \frac{\text{lbmol}}{\text{hr}}$$

Because no materials are added or transformed between the separator and the reactor, the mass flow rate of N leaving the reactor equals the mass flow rate of N entering the separator. The molar flow rate of N entering the separator is

$$F_{N_S} = F_{N_R} + 199.98\ \frac{\text{lbmol}}{\text{hr}}$$

The molar flow rate of N in the recycle stream is 2% of the total N entering the separator.

$$F_{N_R} = 0.02 F_{N_S}$$

Combining the two equations gives

$$F_{N_R} = (0.02)\left(F_{N_R} + 199.98\ \frac{\text{lbmol}}{\text{hr}}\right)$$

$$= \frac{(0.02)\left(199.98\ \frac{\text{lbmol}}{\text{hr}}\right)}{1 - 0.02}$$

$$= 4.1\ \text{lbmol/hr} \quad (4\ \text{lbmol/hr})$$

The answer is (C).

Why Other Options Are Wrong

(A) This answer erroneously reports the molar flow rate of M in the product instead of the molar flow rate of N in the recycle stream.

(B) This answer erroneously assumes that the molar flow rate of M out of the reactor equals the molar flow rate of N in the recycle stream.

(D) This answer erroneously reports the molar flow rate of N to the separator instead of the molar flow rate of N in the recycle stream.

SOLUTION 7

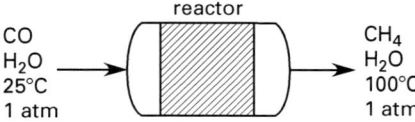

From the balanced reaction, the stoichiometric coefficients are

species	coefficient, σ
CO	1
H_2	3
CH_4	1
H_2O	1

The universal gas constant, R, is 1.987 cal/mol·K. The standard heats of formation and the free energies of formation are given at a temperature of

$$T_{298} = 25°C + 273° = 298K$$

At 25°C, the enthalpy change for the reactants is
$$\Delta H^0_{\text{reactants}} = \sigma_{\text{CO}}\Delta H^0_{f,\text{CO}} + \sigma_{\text{H}_2}\Delta H^0_{f,\text{H}_2}$$
$$= (1)\left(-26\,398\,\frac{\text{cal}}{\text{mol}}\right) + (3)\left(0\,\frac{\text{cal}}{\text{mol}}\right)$$
$$= -26\,398 \text{ cal/mol}$$

At 25°C, the enthalpy change for the products is
$$\Delta H^0_{\text{products}} = \sigma_{\text{CH}_4}\Delta H^0_{f,\text{CH}_4} + \sigma_{\text{H}_2\text{O}}\Delta H^0_{f,\text{H}_2\text{O}}$$
$$= (1)\left(-17\,798\,\frac{\text{cal}}{\text{mol}}\right) + \left(-57\,757\,\frac{\text{cal}}{\text{mol}}\right)$$
$$= -75\,555 \text{ cal/mol}$$

At 25°C, the enthalpy change for the reaction is
$$\Delta H^0_{298} = \Delta H^0_{\text{products}} - \Delta H^0_{\text{reactants}}$$
$$= -75\,555\,\frac{\text{cal}}{\text{mol}} - \left(-26\,398\,\frac{\text{cal}}{\text{mol}}\right)$$
$$= -49\,157 \text{ cal/mol}$$

At 25°C, the change in free-energy of formation for the reactants is
$$\Delta G_{\text{reactants}} = \sigma_{\text{CO}}\Delta G^0_{f,\text{CO}} + \sigma_{\text{H}_2}\Delta G^0_{f,\text{H}_2}$$
$$= (1)\left(-32\,762\,\frac{\text{cal}}{\text{mol}}\right) + (3)\left(0\,\frac{\text{cal}}{\text{mol}}\right)$$
$$= -32\,762 \text{ cal/mol}$$

At 25°C, the change in free-energy of formation for the products is
$$\Delta G_{\text{products}} = \Delta G_{f,\text{H}_2\text{O}} + \Delta G_{f,\text{CH}_4}$$
$$= -54\,593\,\frac{\text{cal}}{\text{mol}} + -12\,052\,\frac{\text{cal}}{\text{mol}}$$
$$= -66\,645 \text{ cal/mol}$$

At 25°C, the free-energy change for the reaction is
$$\Delta G^0 = \Delta G_{\text{products}} - \Delta G_{\text{reactants}}$$
$$= -66\,645\,\frac{\text{cal}}{\text{mol}} - \left(-32\,762\,\frac{\text{cal}}{\text{mol}}\right)$$
$$= -33\,883 \text{ cal/mol}$$

At 298K, the numerical value of the reaction constant is
$$\ln K_{\text{rxn},298} = -\frac{\Delta G^0}{RT} = -\frac{-33\,883\,\frac{\text{cal}}{\text{mol}}}{\left(1.987\,\frac{\text{cal}}{\text{mol}\cdot\text{K}}\right)(298\text{K})}$$
$$= 57.223$$
$$K_{\text{rxn},298} = 7.10 \times 10^{24}$$

At 100°C, the absolute temperature is
$$T_{373} = 100°\text{C} + 273°$$
$$= 373\text{K}$$

For the temperature change from 298K to 373K, the numerical value of the reaction constant is
$$\ln\frac{K_{\text{rxn},373}}{K_{\text{rxn},298}} = \left(\frac{-\Delta H^0_{298}}{R}\right)\left(\frac{1}{T_{373}} - \frac{1}{T_{298}}\right)$$
$$\ln\frac{K_{\text{rxn},373}}{7.10 \times 10^{24}} = \frac{-\left(-49\,157\,\frac{\text{cal}}{\text{mol}}\right)\left(\frac{1}{373\text{K}} - \frac{1}{298\text{K}}\right)}{1.987\,\frac{\text{cal}}{\text{mol}\cdot\text{K}}}$$
$$= -16.693$$

Solving the preceding equation for $K_{\text{rxn},373}$ gives
$$K_{\text{rxn},373} = (7.10 \times 10^{24})e^{-16.693}$$
$$= 3.996 \times 10^{17} \quad (4 \times 10^{17})$$

The answer is (B).

Why Other Options Are Wrong

(A) This incorrect answer uses the wrong value (10.53 cal/mol·°R) of the universal gas constant when calculating the reaction constants.

(C) This incorrect answer exchanges the temperatures when calculating the equilibrium reaction constant.

(D) This answer erroneously adds the free-energy change of carbon monoxide instead of subtracting it when calculating the free-energy change for the reaction.

SOLUTION 8

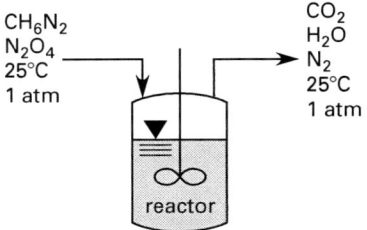

The balanced equation is
$$4\text{CH}_6\text{N}_2(l) + 5\text{N}_2\text{O}_4(l)$$
$$\rightarrow 4\text{CO}_2(g) + 12\text{H}_2\text{O}(g) + 9\text{N}_2(g)$$

From the balanced reaction, the stoichiometric coefficients are

species	coefficient, σ
CH_6N_2	4
N_2O_4	5
CO_2	4
H_2O	12
N_2	9

The enthalpy change and entropy change for the reaction take place at a temperature of

$$T = 25°C + 273°$$
$$= 298 K$$

At 25°C, the enthalpy change for the reactants is

$$\Delta H_{\text{reactants}} = \sigma_{CH_6N_2}\Delta H^0_{f,CH_6N_2} + \sigma_{N_2O_4}\Delta H^0_{f,N_2O_4}$$
$$= (4)\left(54.14\ \frac{kJ}{mol}\right) + (5)\left(9.08\ \frac{kJ}{mol}\right)$$
$$= 261.96\ kJ/mol$$

At 25°C, the enthalpy change for the products is

$$\Delta H^0_{\text{products}} = \sigma_{CO_2}\Delta H^0_{f,CO_2} + \sigma_{H_2O}\Delta H^0_{f,H_2O}$$
$$+ \sigma_{N_2}\Delta H_{N_2}$$
$$= (4)\left(-393.51\ \frac{kJ}{mol}\right)$$
$$+ (12)\left(-241.83\ \frac{kJ}{mol}\right) + (9)\left(0\ \frac{kJ}{mol}\right)$$
$$= -4476\ kJ/mol$$

At 25°C, the enthalpy change for the reaction is

$$\Delta H^0_{298} = \Delta H^0_{\text{products}} - \Delta H^0_{\text{reactants}}$$
$$= -4476\ \frac{kJ}{mol} - 261.96\ \frac{kJ}{mol}$$
$$= -4737.96\ kJ/mol$$

At 25°C, the change in entropy for the reactants is

$$\Delta S_{\text{reactants}} = \sigma_{CH_6N_2}S^0_{CH_6N_2} + \sigma_{N_2O_4}S^0_{N_2O_4}$$
$$= (4)\left(165.94\ \frac{kJ}{mol\cdot K}\right)$$
$$+ (5)\left(304.38\ \frac{J}{mol\cdot K}\right)$$
$$= 2185.66\ J/mol\cdot K$$

At 25°C, the change in entropy for the products is

$$\Delta S_{\text{products}} = \sigma_{CO_2}S^0_{CO_2} + \sigma_{H_2O}S^0_{H_2O} + \sigma_{N_2}S^0_{N_2}$$
$$= (4)\left(213.785\ \frac{J}{mol\cdot K}\right)$$
$$+ (12)\left(188.835\ \frac{J}{mol\cdot K}\right)$$
$$+ (9)\left(191.56\ \frac{J}{mol\cdot K}\right)$$
$$= 4845.2\ J/mol\cdot K$$

At 25°C, the entropy change for the reaction is

$$\Delta S^0_{298} = \Delta S_{\text{products}} - \Delta S_{\text{reactants}}$$
$$= 4845.2\ \frac{J}{mol\cdot K} - 2185.66\ \frac{J}{mol\cdot K}$$
$$= 2659.54\ J/mol\cdot K$$

At 298.15K, the Gibbs free energy is

$$\Delta G^0_{298} = \Delta H^0_{298} - T\Delta S^0_{298}$$
$$= \left(-4737.96\ \frac{kJ}{mol}\right)\left(1000\ \frac{J}{kJ}\right)$$
$$- (298K)\left(2659.5\ \frac{J}{mol\cdot K}\right)$$
$$= -5\,530\,889.9\ J/mol\quad (-5500\ kJ/mol)$$

The answer is (A).

Why Other Options Are Wrong

(B) This incorrect answer uses the temperature in degrees Celsius instead of the absolute scale.

(C) This incorrect answer adds the second term in the expression for the Gibbs free energy instead of subtracting it.

(D) This incorrect answer omits the conversion factor from kilojoules to joules while using the enthalpy in kilojoules per mole and the entropy in joules per mole kelvin.

SOLUTION 9

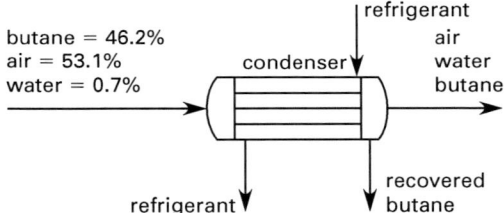

The removal efficiency, η, is 0.85.

Because the butane in the feed is 46.2% (by mole), the butane inlet volume fraction, y, is 0.462. The molecular weight of butane, MW, is given as 58.08 lbm/lbmol. Because the process is at standard conditions, the total pressure, p, is 14.7 lbf/in². The temperature is

$$T = 77°F + 460°$$
$$= 537°R$$

The universal gas constant, R, is 10.73 lbf-ft³/lbmol-in²-°R. The molar specific volume of the inlet stream is

$$v = \frac{RT}{p} = \frac{\left(10.73\ \frac{lbf\text{-}ft^3}{lbmol\text{-}in^2\text{-}°R}\right)(537°R)}{14.7\ \frac{lbf}{in^2}}$$
$$= 392\ ft^3/lbmol$$

The inlet stream flow rate, $\dot{V}$, is given as 300 ft³/min. The molar flow rate of butane in the inlet stream is

$$F = \left(\frac{\dot{V}}{v}\right) y = \left(\frac{300 \ \frac{\text{ft}^3}{\text{min}}}{392 \ \frac{\text{ft}^3}{\text{lbmol}}}\right) (0.462) \left(60 \ \frac{\text{min}}{\text{hr}}\right)$$

$$= 21.21 \ \text{lbmol/hr}$$

The molar flow rate of butane in the outlet stream is

$$F_{\text{out}} = F(1-\eta) = \left(21.21 \ \frac{\text{lbmol}}{\text{hr}}\right)(1-0.85)$$

$$= 3.18 \ \text{lbmol/hr}$$

The recovered condensed butane molar flow rate is

$$F_{\text{cond}} = F - F_{\text{out}}$$
$$= 21.21 \ \frac{\text{lbmol}}{\text{hr}} - 3.18 \ \frac{\text{lbmol}}{\text{hr}}$$
$$= 18.03 \ \text{lbmol/hr}$$

The mass flow rate of butane recovered is

$$\dot{m}_{\text{C}_4\text{H}_{10}} = F_{\text{cond}}(\text{MW})$$
$$= \left(18.03 \ \frac{\text{lbmol}}{\text{hr}}\right)\left(58.08 \ \frac{\text{lbm}}{\text{lbmol}}\right)$$
$$= 1047.18 \ \text{lbm/hr} \quad (1050 \ \text{lbm/hr})$$

The answer is (C).

Why Other Options Are Wrong

(A) This incorrect answer omits the conversion factor from minutes to hours.

(B) This incorrect answer omits the multiplication by the molecular weight of the butane when calculating the molar flow rate of butane. This answer reports the molar flow rate of butane condensed as the answer.

(D) This incorrect answer does not calculate the butane condensed. This answer reports the molar mass flow rate of butane in the inlet stream as the answer.

SOLUTION 10

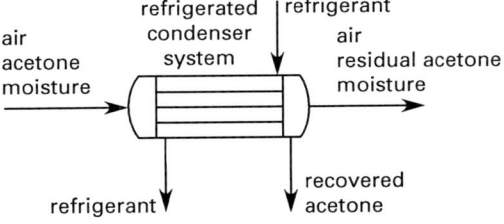

The removal efficiency, η, is given as 0.90. The acetone inlet volume fraction, y_{ac}, is given as 0.385. The total pressure, p, is given as 760 mm Hg. The partial pressure of acetone at the outlet of the condenser is

$$p_{\text{ac}} = p \left(\frac{y_{\text{ac}}(1-\eta)}{1-\eta y_{\text{ac}}}\right)$$

$$= (760 \ \text{mm Hg}) \left(\frac{(0.385)(1-0.90)}{1-(0.90)(0.385)}\right)$$

$$= 44.77 \ \text{mm Hg}$$

Solving the Antoine equation for temperature gives the condensation temperature of the acetone. The condensation temperature of the acetone is

$$T = \frac{1210.595}{7.117 - \log p_{\text{ac}}} - 229.664$$

$$= \frac{1210.595}{7.117 - \log 44.77} - 229.664$$

$$= (-8.19°\text{C})\left(\frac{9°\text{C}}{5°\text{C}}\right) + 32°\text{F}$$

$$= 17.3°\text{F} \quad (17°\text{F})$$

The answer is (C).

Why Other Options Are Wrong

(A) This incorrect answer is calculated using the standard pressure in pounds per square inch instead of millimeters of mercury.

(B) This incorrect answer does not convert the temperature from degrees Celsius to degrees Fahrenheit and reports the answer in the wrong units.

(D) This incorrect answer is the temperature calculated for pure acetone. This answer uses the pressure of acetone equal to the total pressure.

SOLUTION 11

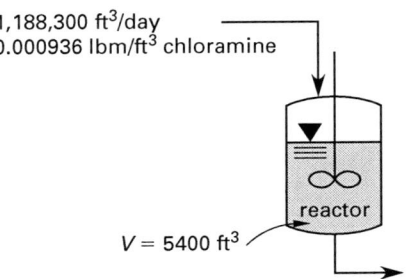

The volumetric liquid flow rate is

$$Q = \left(1{,}188{,}300 \ \frac{\text{ft}^3}{\text{day}}\right)\left(\frac{1 \ \text{day}}{1440 \ \text{min}}\right)$$

$$= 825.2 \ \text{ft}^3/\text{min}$$

Because the average period required to completely renew the fluid in the reactor is the hydraulic residence time, the hydraulic residence time in the tank is

$$\theta = \frac{V}{Q} = \frac{5400 \text{ ft}^3}{825.2 \dfrac{\text{ft}^3}{\text{min}}}$$

$$= 6.54 \text{ min}$$

The influent chloramine concentration is C_i. The output concentration of the chloramine in the tank is C. The mole balance of chloramine for a first-order reaction is

$$\frac{dVC}{dt} = QC_i - QC - kCV$$

Assuming constant volume,

$$V\frac{dC}{dt} = QC_i - QC - kCV$$

Rearranging,

$$\frac{dC}{dt} = \frac{QC_i - QC - kCV}{V}$$

$$= \left(\frac{Q}{V}\right)C_i - \left(\frac{Q}{V}\right)C - kC$$

The residence time is defined as

$$\theta \equiv \frac{V}{Q}$$

The differential equation and initial conditions, where $C = 0$, are

$$\frac{dC}{dt} = \frac{C_i}{\theta} - \left(\frac{1}{\theta} + k\right)C \quad [t = 0]$$

Separating variables gives

$$\int_0^C \frac{dC}{C_i - (1 + k\theta)C} = \frac{1}{\theta}\int_0^t dt$$

Integrating,

$$-\frac{\ln\left(1 - \left(\dfrac{C}{C_i}\right)(1 + k\theta)\right)}{1 + k\theta} = \frac{t}{\theta}$$

Solving for C gives

$$C = C_i\left(\frac{1}{1 + k\theta}\right)\left(1 - e^{-(1+k\theta)(t/\theta)}\right)$$

Because the reaction rate is slow, k may be assumed to be negligible.

$$k \approx 0$$

$$C = C_i\left(1 - e^{-t/\theta}\right)$$

Replacing the chloramine concentration at the inlet condition, given as $C_i = 0.000936$ lbm/ft^3, and the value of the residence time in the preceding equation gives

$$C = 0.000936\left(1 - e^{-t/6.54 \text{ min}}\right)$$

The answer is (A).

Why Other Options Are Wrong

(B) This incorrect answer uses days instead of minutes when calculating the hydraulic residence time.

(C) This answer erroneously uses the opposite sign for the exponential term in the integration.

(D) This answer erroneously uses the opposite sign for the inflow concentration in the integration.

SOLUTION 12

This process of energy transfer between the liquid water and the gas stream is adiabatic because there are no heat losses to the surroundings. The heat is transferred from the gas stream entering the humidifier to the water from the water tank as it evaporates. The spray water will evaporate and its temperature will increase to the gas stream exit temperature of 350°F. As the droplets evaporate, they convert sensible heat in the surrounding gas stream to latent heat.

The enthalpy change of the liquid water from 82°F to 212°F is

$$\Delta H_{82 \to 212} = c_{p,l}(T_{212} - T_{82})$$

$$= \left(1 \frac{\text{Btu}}{\text{lbm-°F}}\right)(212°\text{F} - 82°\text{F})$$

$$= 130 \text{ Btu/lbm}$$

The enthalpy change of the liquid water from 212°F to vapor at 212°F equals the heat of vaporization of the water at 212°F. The heat of vaporization of liquid water is given as

$$\lambda_{H_2O} = \Delta H_{212} = 970.3 \text{ Btu/lbm}$$

The enthalpy change of the water vapor from 212°F to 350°F is

$$\Delta H_{212 \to 350} = c_{p,v}(T_{350} - T_{212})$$

$$= \left(0.48 \frac{\text{Btu}}{\text{lbm-°F}}\right)(350°\text{F} - 212°\text{F})$$

$$= 66.24 \text{ Btu/lbm}$$

The enthalpy change of the water from 82°F to 350°F is

$$\Delta H_{82 \to 350} = \Delta H_{82 \to 212} + \Delta H_{212} + \Delta H_{212 \to 350}$$
$$= 130 \, \frac{\text{Btu}}{\text{lbm}} + 970.3 \, \frac{\text{Btu}}{\text{lbm}} + 66.24 \, \frac{\text{Btu}}{\text{lbm}}$$
$$= 1166.54 \, \text{Btu/lbm}$$

Let the inlet conditions be designated as 1 and the measured conditions be designated as 2. The temperature of the hot gas in is

$$T_1 = T_{\text{in}} + 460° = 800°\text{F} + 460°$$
$$= 1260°\text{R}$$

The temperature at which the gas volume was measured is

$$T_2 = 68°\text{F} + 460° = 528°\text{R}$$

The pressure at which the gas volume was measured, p_2, is 407 in H_2O. For an ideal gas at inlet conditions 1, the equation of state is

$$p_1 \dot{V}_1 = RT_1$$

For an ideal gas at measured conditions 2, the equation of state is

$$p_2 \dot{V}_1 = RT_2$$

Dividing the equation of state at inlet conditions 1 by the equation of state at measured conditions 2 and solving, the volume of the gas at inlet conditions is

$$\dot{V}_1 = \frac{V_2 T_1 p_2}{T_2 p_1}$$
$$= \frac{\left(4000 \, \frac{\text{ft}^3}{\text{min}}\right)(1260°\text{R})(407 \, \text{in} \, H_2O)}{(528°\text{R})(390 \, \text{in} \, H_2O)}$$
$$= 9961.54 \, \text{ft}^3/\text{min}$$

The gas mass flow rate is

$$\dot{m}_{g,\text{inlet}} = \frac{\dot{V}_1 (\text{MW}_g)}{v}$$
$$= \frac{\left(9961.54 \, \frac{\text{ft}^3}{\text{min}}\right)\left(29 \, \frac{\text{lbm}}{\text{lbmol}}\right)}{960 \, \frac{\text{ft}^3}{\text{lbmol}}}$$
$$= 300.92 \, \text{lbm/min}$$

The gas stream is cooled from 800°F to 350°F. The rate of enthalpy change of the gas stream at 800°F is

$$\Delta \dot{H}_{g,800} = \dot{m}_{g,\text{inlet}} H_{g,800}$$
$$= \left(300.92 \, \frac{\text{lbm}}{\text{min}}\right)\left(230.1 \, \frac{\text{Btu}}{\text{lbm}}\right)$$
$$= 69{,}241.69 \, \text{Btu/min}$$

The rate of enthalpy change of the gas stream at 350°F is

$$\Delta \dot{H}_{g,350} = \dot{m}_{g,\text{inlet}} H_{g,350}$$
$$= \left(300.92 \, \frac{\text{lbm}}{\text{min}}\right)\left(62.3 \, \frac{\text{Btu}}{\text{lbm}}\right)$$
$$= 18{,}747.32 \, \text{Btu/min}$$

The rate of enthalpy change of the gas stream as it cools from 800°F to 350°F is

$$\Delta \dot{H}_{g,800 \to 350} = \Delta \dot{H}_{g,800} - \Delta \dot{H}_{g,350}$$
$$= 69{,}241.69 \, \frac{\text{Btu}}{\text{min}} - 18{,}747.32 \, \frac{\text{Btu}}{\text{min}}$$
$$= 50{,}494.38 \, \text{Btu/min}$$

The mass flow rate of water required to cool the gas stream from 800°F to 350°F is

$$\dot{m}_w = \frac{\Delta \dot{H}_{g,800 \to 350}}{\Delta H_{82 \to 350}} = \frac{50{,}494.38 \, \frac{\text{Btu}}{\text{min}}}{1166.54 \, \frac{\text{Btu}}{\text{lbm}}}$$
$$= 43.29 \, \text{lbm/min}$$

The volumetric flow rate of water needed is

$$\dot{V}_{H_2O,\text{needed}} = \frac{\dot{m}_w}{\rho} = \left(\frac{43.29 \, \frac{\text{lbm}}{\text{min}}}{62.19 \, \frac{\text{lbm}}{\text{ft}^3}}\right)\left(60 \, \frac{\text{min}}{\text{hr}}\right)$$
$$= 41.77 \, \text{ft}^3/\text{hr}$$

The volume of the water tank, V, is given as 2036 ft^3. Because the water tank is 70% filled, the volume of the water in the water tank is

$$V_{H_2O} = 0.7V = (0.7)(2036 \, \text{ft}^3)$$
$$= 1425.2 \, \text{ft}^3$$

The length of time during which the water in the tank (without replenishment) will be able to cool the gas stream is

$$t = \frac{V_{H_2O}}{\dot{V}_{H_2O,\text{needed}}} = \frac{1425.2 \, \text{ft}^3}{41.77 \, \frac{\text{ft}^3}{\text{hr}}}$$
$$= 34.14 \, \text{hr} \quad (34 \, \text{hr})$$

The answer is (A).

Why Other Options Are Wrong

(B) This incorrect answer uses the tank volume instead of the water volume when calculating the time.

(C) This answer erroneously uses the rate of enthalpy change of the gas stream at 350°F instead of the enthalpy change of the gas stream as it cools from 800°F to 350°F.

(D) This answer erroneously omits the factor to convert from minutes to hours.

SOLUTION 13

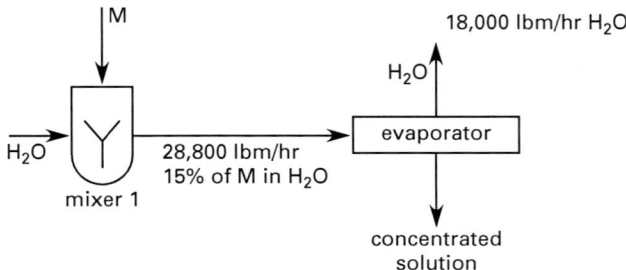

The mass flow rate of feed solution to the evaporator, $\dot{m}$, is given as 28,800 lbm/hr. The mass flow rate of M in the solution is

$$\dot{m}_M = (15\%)\dot{m}$$
$$= \left(0.15 \, \frac{\text{lbm M}}{\text{lbm solution}}\right)\left(28{,}800 \, \frac{\text{lbm solution}}{\text{hr}}\right)$$
$$= 4320 \text{ lbm M/hr}$$

The mass flow rate of water in the solution is

$$\dot{m}_{H_2O} = (100\% - 15\%)\dot{m}$$
$$= \left(0.85 \, \frac{\text{lbm } H_2O}{\text{lbm solution}}\right)\left(28{,}800 \, \frac{\text{lbm solution}}{\text{hr}}\right)$$
$$= 24{,}480 \text{ lbm } H_2O/\text{hr}$$

The molecular weight of M, MW_M, is given as 42.39 lbm M/lbmol M. The molar flow rate of M in the solution is

$$F_M = \frac{\dot{m}_M}{MW_M} = \frac{4320 \, \frac{\text{lbm M}}{\text{hr}}}{42.39 \, \frac{\text{lbm M}}{\text{lbmol M}}}$$
$$= 101.91 \text{ lbmol M/hr}$$

The molecular weight of water, MW_{H_2O}, is given as 18.015 lbm H_2O/lbmol H_2O. The molar flow rate of water in the solution is

$$F_{H_2O} = \frac{\dot{m}_{H_2O}}{MW_{H_2O}} = \frac{24{,}480 \, \frac{\text{lbm } H_2O}{\text{hr}}}{18.015 \, \frac{\text{lbm } H_2O}{\text{lbmol } H_2O}}$$
$$= 1358.87 \text{ lbmol } H_2O/\text{hr}$$

The ratio of the molar flow rate of water to the molar flow rate of M is

$$\text{ratio} = \frac{F_{H_2O}}{F_M} = \frac{1358.87 \, \frac{\text{lbmol } H_2O}{\text{hr}}}{101.91 \, \frac{\text{lbmol M}}{\text{hr}}}$$
$$= 13.33 \text{ lbmol } H_2O/\text{lbmol M}$$

Starting with pure M and pure H_2O, the process of forming the solution is

$$M + H_2O \rightarrow \text{solution}$$

The enthalpy change of the process is equal to the rate of heat removed in the process, Q.

$$\Delta H_{\text{sol,tank}} - \Delta H_M - \Delta H_{H_2O} = Q$$

By convention, the enthalpies of the pure components are zero at 77°F. The enthalpies of pure M and pure H_2O are zero. The enthalpy of the solution in the tank is equal to the heat removed when preparing the solution of pure M in pure water. At a molar ratio of 13.33 and 77°F, from the data given, the enthalpy of the solution per mole of M, $\Delta H_{13.33}$, is $-32{,}036$ Btu/lbmol. The enthalpy of the solution formed is

$$\Delta H_{\text{sol,tank}} = \Delta H_{13.33} F_M$$
$$= \left(-32{,}036 \, \frac{\text{Btu}}{\text{lbmol M}}\right)\left(101.91 \, \frac{\text{lbmol M}}{\text{hr}}\right)$$
$$= -3.3 \times 10^6 \text{ Btu/hr}$$

The minus sign indicates that the process of dissolving M in water is an exothermic process. Because the rate of heat removed in the process equals the enthalpy change of the process, the rate of heat removed in the process is

$$Q = -3.3 \times 10^6 \, \frac{\text{Btu}}{\text{hr}} - 0 \, \frac{\text{Btu}}{\text{hr}} - 0 \, \frac{\text{Btu}}{\text{hr}}$$
$$= -3.3 \times 10^6 \text{ Btu/hr}$$

Therefore, the rate of heat produced, Q, is 3.3×10^6 Btu/hr.

The answer is (D).

Why Other Options Are Wrong

(A) This incorrect answer does not recognize the enthalpy change when preparing a solution of M in water.

(B) This answer erroneously calculates the rate of heat removed with the solution out of the evaporator instead of the rate of heat removed with the solution formed.

(C) This answer erroneously uses the mass ratio instead of the mole ratio.

SOLUTION 14

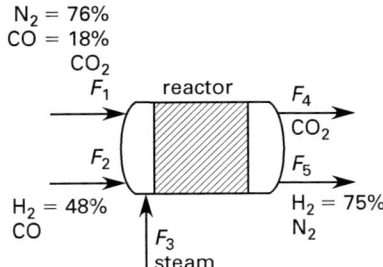

The molar flow rate of stream 1, F_1, is given as 100 lbmol/hr. Because the mole fraction of N_2 is 0.76 in stream 1 and 0.25 in stream 5, a mole balance of N_2 around the reactor gives the molar flow rate of stream 5. The N_2 mole balance around the reactor gives

$$0.25 F_5 = 0.76 F_1$$
$$= (0.76)\left(100 \ \frac{\text{lbmol}}{\text{hr}}\right)$$
$$= 76 \ \text{lbmol/hr}$$

Solving the preceding equation for F_5 gives

$$F_5 = \frac{76 \ \frac{\text{lbmol}}{\text{hr}}}{0.25} = 304 \ \text{lbmol/hr}$$

Because stream 5 consists of 75 mol% H_2, the fraction of H_2 in stream 5, $x_{H_2,5}$, is 0.75. The molar flow rate of H_2 in stream 5 is

$$F_{H_2,5} = x_{H_2,5} F_5 = (0.75)\left(304 \ \frac{\text{lbmol}}{\text{hr}}\right)$$
$$= 228 \ \text{lbmol/hr}$$

Because stream 2 consists of 48 mol% H_2, the fraction of H_2 in stream 2, $x_{H_2,2}$, is 0.48. Because stream 2 contains only H_2 and CO,

$$x_{H_2,2} + x_{CO,2} = 1$$
$$0.48 + x_{CO,2} = 1$$

The mole fraction of CO in stream 1, $x_{CO,1}$, is 0.18. The mole fraction of CO in stream 2 is

$$x_{CO,2} = 1 - x_{H_2,2}$$
$$= 1 - 0.48$$
$$= 0.52$$

According to the balanced reaction $CO + H_2O \rightarrow CO_2 + H_2$, the number of moles per hour of CO that reacts produces the same number of moles of H_2. The number of moles of CO per hour that reacted is r. Because the CO reacted completely, the mole balance of CO around the reactor gives

$$0 \ \frac{\text{lbmol}}{\text{hr}} = x_{CO,1} F_1 + x_{CO,2} F_2 - r$$

Solving for r gives

$$r = x_{CO,1} F_1 + x_{CO,2} F_2 \qquad \text{[I]}$$

A mole balance of H_2 around the reactor states that the H_2 in stream 5 equals the H_2 produced from the reaction of CO in streams 1 and 2 plus the H_2 fed in stream 2. Because the mole fraction of H_2 in stream 2 is 0.48, the mole balance of H_2 around the reactor is

$$x_{H_2,5} F_5 = x_{H_2,2} F_2 + r \qquad \text{[II]}$$

Replacing Eq. I in Eq. II gives

$$x_{H_2,5} F_5 = x_{H_2,2} F_2 + x_{CO,1} F_1 + x_{CO,2} F_2$$

Solving,

$$F_2 = \frac{x_{H_2,5} F_5 + x_{CO,1} F_1}{x_{H_2,2} + x_{CO,2}}$$
$$= \frac{(0.75)\left(304 \ \frac{\text{lbm}}{\text{hr}}\right) - (0.18)\left(100 \ \frac{\text{lbm}}{\text{hr}}\right)}{0.48 + 0.52}$$
$$= 210 \ \text{lbm/hr}$$

Replacing F_2 in Eq. I gives

$$r = (0.18)\left(100 \ \frac{\text{lbm}}{\text{hr}}\right) + (0.52)\left(210 \ \frac{\text{lbm}}{\text{hr}}\right)$$
$$= 127.2 \ \text{lbm/hr}$$

A mole balance of water around the reactor gives

$$0 \ \frac{\text{lbmol}}{\text{hr}} = F_3 - r$$
$$F_3 = 127.2 \ \text{mol/hr}$$

The ratio of stream 1 to steam stream 3, is

$$\frac{F_1}{F_3} = \frac{100 \ \frac{\text{lbmol}}{\text{hr}}}{127.2 \ \frac{\text{lbmol}}{\text{hr}}}$$
$$= 0.786 \quad (0.79)$$

The answer is (C).

Why Other Options Are Wrong

(A) This incorrect answer is the ratio of the molar flow rates of streams 1 and 5 instead of the ratio of streams 1 and 3.

(B) This incorrect answer is the ratio of the molar flow rates of streams 1 and 2 instead of the ratio of streams 1 and 3.

(D) This incorrect answer is the inverse of the ratio of the molar flow rates of streams 1 and 3; in other words, this answer is the ratio of streams 3 and 1.

SOLUTION 15

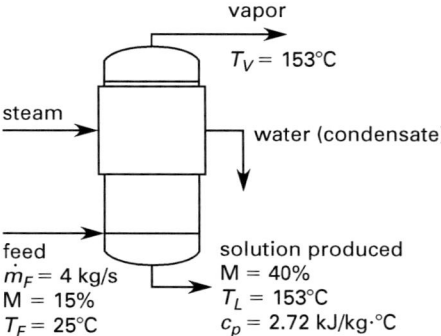

The mass flow rate of the feed solution to the evaporator, $\dot{m}_F$, is given as 4 kg/s. Because the feed solution is 15% M, the mass flow rate of M in the feed is

$$\dot{m}_M = (15\%)\dot{m}_F$$
$$= \left(0.15 \; \frac{\text{kg M}}{\text{kg feed}}\right)\left(4.0 \; \frac{\text{kg feed}}{\text{s}}\right)$$
$$= 0.6 \; \text{kg M/s}$$

Because the solution produced by the evaporator is 40% M, the mass flow rate of the solution produced by the evaporator is

$$\dot{m}_L = \frac{0.6 \; \frac{\text{kg M}}{\text{s}}}{0.40 \; \frac{\text{kg M}}{\text{kg solution}}}$$
$$= 1.50 \; \text{kg solution/s}$$

The molecular weight of M, MW_M, is given as 42.39 kg/kmol. The molar flow rate of M in the feed solution is

$$F_{M_F} = \frac{\dot{m}_M}{MW_M} = \frac{\left(0.6 \; \frac{\text{kg M}}{\text{s}}\right)\left(1000 \; \frac{\text{g}}{\text{kg}}\right)}{42.39 \; \frac{\text{g}}{\text{mol M}}}$$
$$= 14.15 \; \text{mol M/s}$$

Calculate the enthalpy change of the solution produced in the evaporator when the solution is heated from 25°C to 153°C. The total enthalpy change is the total enthalpy of the product minus the total enthalpy of the feed stream. Take the reference temperature to be that of the feed stream, 25°C. The reference state is liquid.

The heat of solution of the solution fed into the evaporator, ΔH_F, is given as -33.810 kJ/mol M. The rate of enthalpy change of the solution fed into the evaporator is

$$\Delta H_a = \Delta H_F F_{M_F}$$
$$= \left(-33.810 \; \frac{\text{kJ}}{\text{mol M}}\right)\left(14.15 \; \frac{\text{mol M}}{\text{s}}\right)$$
$$= -478.41 \; \text{kJ/s}$$

The enthalpy of the feed stream involves the separation of 4 kg of 15% M solution into its pure constituents at 25°C. The heat effect of the corresponding mixing process is identical in magnitude to the process of separating M and H_2O, but with opposite sign.

The rate of heat transfer in this step equals the rate of enthalpy change of the solution fed into the evaporator.

$$Q_F = -\Delta H_a = 478.41 \; \text{kJ/s}$$

Notice there was no specific heat contribution in the feed because the reference temperature is the feed temperature. The heat of solution of the solution produced by the evaporator, ΔH_L, is given as -23.267 kJ/mol M. The rate of enthalpy change of the solution produced by the evaporator is

$$\Delta H_b = \Delta H_L F_{M_F}$$
$$= \left(-23.267 \; \frac{\text{kJ}}{\text{mol M}}\right)\left(14.15 \; \frac{\text{mol M}}{\text{s}}\right)$$
$$= -329.228 \; \text{kJ/s}$$

The rate of heat transfer equals the rate of enthalpy change of the solution produced by the evaporator.

$$Q_L = \Delta H_b = -329.23 \; \text{kJ/s}$$

The heat capacity of the liquid solution from the evaporation, c_p, is given as 2.72 kJ/kg solution·°C. The mass flow rate of M in the solution produced in the evaporator, $\dot{m}_{L_M}$, is given as 0.60 kg/s. The rate of heat absorbed by the solution produced in the evaporator as it goes from 25°C to 153°C is

$$Q_{\text{sol},25\to153} = \dot{m}c_p\Delta T = \dot{m}_L c_p(T_L - T_F)$$
$$= \left(1.50 \; \frac{\text{kg solution}}{\text{s}}\right)$$
$$\times \left(2.72 \; \frac{\text{kJ}}{\text{kg solution·°C}}\right)(153°\text{C} - 25°\text{C})$$
$$= 522.24 \; \text{kJ/s}$$

The calculation of the enthalpy change when liquid water is vaporized from the solution and superheated at 153°C follows.

The enthalpy of saturated steam at 153°C, $\Delta H_{g,153}$, is given as 2750 kJ/kg. The enthalpy of liquid water at 25°C and 1 atm, $\Delta H_{f,25}$, is given as 104.89 kJ/kg. The rate of heat absorbed by the saturated vapor at 153°C is

$$Q_{V,153} = \dot{m}_{V_{H_2O}}(\Delta H_{g,153} - \Delta H_{f,25})$$
$$= \left(2.50 \ \frac{\text{kg H}_2\text{O}}{\text{s}}\right)$$
$$\times \left(2750 \ \frac{\text{kJ}}{\text{kg H}_2\text{O}} - 104.89 \ \frac{\text{kJ}}{\text{kg H}_2\text{O}}\right)$$
$$= 6612.775 \text{ kJ/s}$$

The total heat balance around the evaporator is

$$Q_S = Q_F + Q_L + Q_{\text{sol},25 \to 153} + Q_{V,153}$$
$$= 478.41 \ \frac{\text{kJ}}{\text{s}} + \left(-329.23 \ \frac{\text{kJ}}{\text{s}}\right)$$
$$+ 522.24 \ \frac{\text{kJ}}{\text{s}} + 6612.775 \ \frac{\text{kJ}}{\text{s}}$$
$$= 7284.195 \text{ kJ/s}$$

The latent heat, λ, is given as 1715.3 kJ/kg. The energy from superheating and subcooling is assumed to be negligible compared to the latent heat. When calculating the steam mass flow rate, only the latent heat is to be considered. The steam mass flow rate is

$$\dot{m}_S = \frac{Q_S}{\lambda} = \frac{7284.195 \ \frac{\text{kJ}}{\text{s}}}{1715.3 \ \frac{\text{kJ}}{\text{kg}}}$$
$$= 4.247 \text{ kg/s} \quad (4.2 \text{ kg/s})$$

The answer is (C).

Why Other Options Are Wrong

(A) This incorrect answer omits the rate of heat transfer for the vapor at 153°C to superheated vapor at 153°C.

(B) This incorrect answer calculates the heat transfer using the wrong sign for the second term in the expression of the rate of heat transfer in the evaporator.

(D) This incorrect answer uses the rate of heat produced by the solution in the evaporator with the opposite sign.

SOLUTION 16

The waste gas inlet temperature, $T_{\text{wg,in}}$, is given as 110°F. The incinerator operating temperature, $T_{\text{fg,in}}$, is given as 1600°F. The mass flow rate of the waste gas in the preheater is $\dot{m}_{\text{wg,in}}$. The mass flow rate of the waste gas out of the preheater is $\dot{m}_{\text{wg,out}}$. The heat capacity of the waste gas in the preheater is $c_{p,\text{wg,in}}$. The waste gas enters the preheater at $T_{\text{wg,in}}$ and leaves the preheater at $T_{\text{wg,out}}$. The mass flow rates of the gases on both sides of the preheater are exactly the same, if there are no leaks.

$$\dot{m}_{\text{wg,out}} = \dot{m}_{\text{wg,in}} = \dot{m}_{\text{wg}}$$

The heat capacities of the gases on both sides of the preheater are approximately the same.

$$c_{p,\text{wg,out}} \approx c_{p,\text{wg,in}} \equiv c_{p,\text{wg}}$$

The maximum caloric energy recovery in the preheater occurs when $T_{\text{wg,out}}$ equals $T_{\text{fg,in}}$.

$$\text{EOP} = \dot{m}_{\text{wg,out}} c_{p,\text{wg,out}} T_{\text{wg,out}} - \dot{m}_{\text{wg,in}} c_{p,\text{wg,in}} T_{\text{wg,in}}$$
$$\approx \dot{m}_{\text{wg}} c_{p,\text{wg}} (T_{\text{wg,out}} - T_{\text{wg,in}})$$
$$= \dot{m}_{\text{wg}} c_{p,\text{wg}} (T_{\text{fg,in}} - T_{\text{wg,in}})$$

The energy actually recovered is the increase in sensible heat of the waste gas.

$$\text{CE} = \dot{m}_{\text{wg,out}} c_{p,\text{wg,out}} T_{\text{wg,out}} - \dot{m}_{\text{wg,in}} c_{p,\text{wg,in}} T_{\text{wg,in}}$$
$$\approx \dot{m}_{\text{wg}} c_{p,\text{wg}} (T_{\text{wg,out}} - T_{\text{wg,in}})$$

The fractional energy recovery is given as

$$E_{\text{rec}} = \frac{\text{CE}}{\text{EOP}} = 0.75$$
$$0.75 = \frac{\dot{m}_{\text{wg}} c_{p,\text{wg}} (T_{\text{wg,out}} - T_{\text{wg,in}})}{\dot{m}_{\text{wg}} c_{p,\text{wg}} (T_{\text{fg,in}} - T_{\text{wg,in}})} = \frac{T_{\text{wg,out}} - T_{\text{wg,in}}}{T_{\text{fg,in}} - T_{\text{wg,in}}}$$
$$T_{\text{wg,out}} = 0.75(T_{\text{fg,in}} - T_{\text{wg,in}}) + T_{\text{wg,in}}$$
$$= (0.75)(1600°\text{F} - 110°\text{F}) + 110°\text{F}$$
$$= 1227.5°\text{F}$$

The caloric energy change of the flue gas in the preheater is

$$\Delta E_f = \dot{m}_{\text{wg}} c_{p,\text{wg}} (T_{\text{fg,in}} - T_{\text{fg,out}})$$

The energy balance around the preheater gives the energy change of the flue gas in the preheater, and this energy change equals the energy change of the waste gas in the preheater.

$$\Delta E_f = \dot{m}_{\text{wg}} c_{p,\text{wg}} (T_{\text{fg,in}} - T_{\text{fg,out}})$$
$$= \dot{m}_{\text{wg}} c_{p,\text{wg}} (T_{\text{wg,out}} - T_{\text{wg,in}})$$

Simplifying,

$$T_{fg,in} - T_{fg,out} = T_{wg,out} - T_{wg,in}$$
$$T_{fg,out} = T_{fg,in} - T_{wg,out} + T_{wg,in}$$
$$= 1600°F - 1227.5°F + 110°F$$
$$= 482.5°F \quad (480°F)$$

The answer is (C).

Why Other Options Are Wrong

(A) This incorrect answer is the waste gas inlet temperature. This answer assumes the waste gas does not change in temperature when it goes through the pre-heater.

(B) This incorrect answer adds the temperature of the inlet waste gas instead of subtracting it when calculating the outlet waste gas temperature.

(D) This incorrect answer adds the outlet waste gas temperature instead of subtracting it when calculating the flue gas outlet temperature.

SOLUTION 17

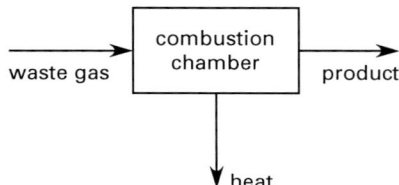

The mole fraction of benzene, $x_{C_6H_6}$, is given as 1000 ppm by volume, which equals 0.001. The mole fraction of chloroform, x_{CHCl_3}, is given as 1000 ppm by volume, which equals 0.001. The molecular weight of the waste gas, MW, is given as 28.97 lbm/lbmol. The temperature is

$$T = 77°F + 460° = 537°R$$

The density of the waste gas at 537°F is

$$\rho = \frac{(MW)p}{RT} = \frac{\left(28.97 \frac{lbm}{lbmol}\right)\left(14.7 \frac{lbf}{in^2}\right)}{\left(10.73 \frac{lbf\text{-}ft^3}{lbmol\text{-}in^2\text{-}°R}\right)(537°R)}$$
$$= 0.0739 \ lbm/ft^3$$

The combustion reaction takes place with air, the third component of the gas stream. The heat of combustion is negative because the reaction is exothermic. The volumetric heat of combustion for C_6H_6, $\Delta H_{C_6H_6}$, is given as -3616 Btu/ft³. The volumetric heat of combustion

for $CHCl_3$, ΔH_{CHCl_3}, is given as -705 Btu/ft³. The volumetric heat of combustion of the gas stream is

$$\Delta H = \Delta H_{C_6H_6} x_{C_6H_6} + \Delta H_{CHCl_3} x_{CHCl_3}$$
$$= \left(-3616 \frac{Btu}{ft^3}\right)(0.001) + \left(-705 \frac{Btu}{ft^3}\right)(0.001)$$
$$= -4.32 \ Btu/ft^3$$

The heat of combustion of the waste gas stream is

$$\Delta H_w = \frac{\Delta H}{\rho} = \frac{-4.32 \frac{Btu}{ft^3}}{0.0739 \frac{lbm}{ft^3}}$$
$$= -58.46 \ Btu/lbm \quad (-58 \ Btu/lbm)$$

The answer is (A).

Why Other Options Are Wrong

(B) This incorrect answer subtracts the heat of combustion of chloroform instead of adding it when calculating the heat of combustion of the gas stream.

(C) This incorrect answer does not calculate the density of the waste stream. This answer reports incorrect units.

(D) This incorrect answer changes the sign of the correct answer. This answer ignores the convention that the heat of combustion is negative.

SOLUTION 18

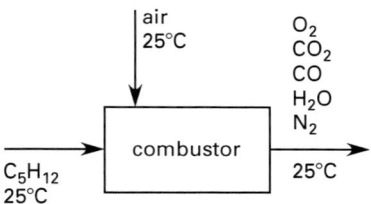

Consider the following reaction. Assume the chemical reaction goes to complete combustion to calculate the amount of oxygen and nitrogen in the dry air supplied. In this reaction, all reactants and products are in the gaseous phase.

$$C_5H_{12} + 8O_2 \rightarrow 5CO_2 + 6H_2O$$

The ratio of nitrogen to oxygen in air is given as 0.79 mol N_2/0.21 mol O_2. The number of moles of CO_2 formed per mole of n-pentane burnt is X. Because the combustion is done with 100% excess air, the number of moles of O_2 used is two times its stoichiometric coefficient.

$$n_{O_2} = (\text{stoichiometric coefficient})(2 \text{ mol})$$
$$= (8)(2 \text{ mol})$$
$$= 16 \text{ mol}$$

From the stoichiometry of the combustion reaction, the number of moles of O_2 in the product is

$$n_{O_2,p} = 16 \text{ mol} - X - \frac{5 \text{ mol} - X}{2} - \frac{6 \text{ mol}}{2}$$
$$= \frac{21 \text{ mol} - X}{2}$$

The balanced equation, with 100% excess air, is as follows.

$$C_5H_{12} + 16O_2 + n_{O_2}\left(\frac{79 \text{ mol}}{21 \text{ mol}}\right) N_2$$
$$\to X CO_2 + (5 \text{ mol} - X)CO + 6 \text{ mol } H_2O + n_{O_2}$$
$$\times \left(\frac{79 \text{ mol}}{21 \text{ mol}}\right) N_2 + n_{O_2,p} O_2$$

Substituting,

$$C_5H_{12} + 16O_2 + (16 \text{ mol})\left(\frac{79 \text{ mol}}{21 \text{ mol}}\right) N_2$$
$$\to X CO_2 + (5 \text{ mol} - X)CO + 6 \text{ mol } H_2O$$
$$+ (16 \text{ mol } O_2)\left(\frac{79 \text{ mol}}{21 \text{ mol}}\right) N_2 + \left(\frac{21 \text{ mol} - X}{2}\right) O_2$$

The number of moles of N_2 in the product is

$$n_{N_2,p} = (16 \text{ mol})\left(\frac{79 \text{ mol}}{21 \text{ mol}}\right)$$
$$= 60.2 \text{ mol}$$

The stoichiometry table for the reaction is

component	reactants (mol)	products (mol)
C_5H_{12}	1	0
O_2	16	$(21 \text{ mol} - X)/2$
CO_2	0	X
CO	0	$5 \text{ mol} - X$
$H_2O(g)$	0	6
N_2	60.2	60.2

The heat of reaction used to heat the products and the excess reactants at 25°C is

$$\Delta H_{\text{rxn}} = X \Delta H_{f,CO_2} + (5 \text{ mol} - X)\Delta H_{f,CO}$$
$$+ (6 \text{ mol})\Delta H_{f,H_2O} - (1 \text{ mol})\Delta H_{f,C_5H_{12}}$$
$$= X\left(-94\,051.8 \, \frac{\text{cal}}{\text{mol}}\right) + (5 \text{ mol} - X)$$
$$\times \left(-26\,415.7 \, \frac{\text{cal}}{\text{mol}}\right) + (6 \text{ mol})$$
$$\times \left(-57\,798 \, \frac{\text{cal}}{\text{mol}}\right)$$
$$- (1 \text{ mol})\left(-30\,086 \, \frac{\text{cal}}{\text{mol}}\right)$$
$$= -\left(448\,780.5 \text{ cal} + \left(67\,636.1 \, \frac{\text{cal}}{\text{mol}}\right)X\right)$$

The gases enter the chamber at

$$T_1 = 25°C + 273° = 298K$$

The gases leave the chamber at

$$T_2 = 827°C + 273° = 1100K$$

The temperature difference is

$$\Delta T = T_2 - T_1 = 1100K - 298K$$
$$= 802K$$

Calculating the enthalpy of each component in the products, the enthalpy change for O_2 when it goes from 25°C to product at 827°C is

$$H_{p_{O_2}} = \overline{C}_{p,O_2} \Delta T = \left(7.82 \, \frac{\text{cal}}{\text{mol·K}}\right)(1100K - 298K)$$
$$= 6271.64 \text{ cal/mol}$$

The enthalpy change for CO_2 when it goes from 25°C to product at 827°C is

$$H_{p_{CO_2}} = \overline{C}_{p,CO_2} \Delta T = \left(7.47 \, \frac{\text{cal}}{\text{mol·K}}\right)(1100K - 298K)$$
$$= 5990.94 \text{ cal/mol}$$

The enthalpy change for CO when it goes from 25°C to product at 827°C is

$$H_{p_{CO}} = \overline{C}_{p,CO} \Delta T = \left(11.61 \, \frac{\text{cal}}{\text{mol·K}}\right)(1100K - 298K)$$
$$= 9311.22 \text{ cal/mol}$$

The enthalpy change for H_2O when it goes from 25°C to product at 827°C is

$$H_{p_{H_2O}} = \overline{C}_{p,H_2O} \Delta T = \left(9.02 \, \frac{\text{cal}}{\text{mol·K}}\right)(1100K - 298K)$$
$$= 7234.04 \text{ cal/mol}$$

The enthalpy change for N_2 when it goes from 25°C to product at 827°C is

$$H_{p_{N_2}} = \overline{C}_{p,N_2} \Delta T = \left(7.39 \, \frac{\text{cal}}{\text{mol·K}}\right)(1100K - 298K)$$
$$= 5926.78 \text{ cal/mol}$$

The energy balance around the combustion chamber gives the total heat transferred.

$$Q = \left(\frac{21 \text{ mol} - X}{2}\right)\Delta H_{p,O_2} + X\Delta H_{p,CO_2}$$
$$+ (5 \text{ mol} - X)\Delta H_{p,CO} + (6 \text{ mol})\Delta H_{p,H_2O}$$
$$+ (60.2 \text{ mol})\Delta H_{p,N_2}$$
$$= \left(\frac{21 \text{ mol} - X}{2}\right)\left(6271.64 \, \frac{\text{cal}}{\text{mol}}\right) + X\left(5990.04 \, \frac{\text{cal}}{\text{mol}}\right)$$
$$+ (5 \text{ mol} - X)\left(9311.22 \, \frac{\text{cal}}{\text{mol}}\right)$$
$$+ (6 \text{ mol})\left(7234.04 \, \frac{\text{cal}}{\text{mol}}\right)$$
$$+ (60.2 \text{ mol})\left(5926.78 \, \frac{\text{cal}}{\text{mol}}\right)$$
$$= 512\,604.7 \text{ cal} - \left(6457 \, \frac{\text{cal}}{\text{mol}}\right) X$$

The total heat transferred equals the negative value of the reaction enthalpy previously calculated as a function of the number of moles of CO_2 formed per mole of n-pentane burnt (because the combustor is adiabatic).

$$448\,780.5 \text{ cal} + \left(67\,636.1 \, \frac{\text{cal}}{\text{mol}}\right) X$$
$$= 512\,604.7 \text{ cal} - \left(6457 \, \frac{\text{cal}}{\text{mol}}\right) X$$

Solving for X gives

$$X = \frac{512\,604.7 \text{ cal} - 448\,780.5 \text{ cal}}{67\,636.1 \, \frac{\text{cal}}{\text{mol}} + 6457 \, \frac{\text{cal}}{\text{mol}}}$$
$$= 0.861 \text{ mol}$$

According to the balanced chemical reaction, the 100% conversion of pentane produces 5 mol of CO_2. The percent of n-pentane burned to CO_2 is

$$\% \text{ n-pentane} = \frac{X \times 100\%}{5 \text{ mol}} = \frac{(0.861 \text{ mol}) \times 100\%}{5 \text{ mol}}$$
$$= 17.2\% \quad (17\%)$$

The answer is (D).

Why Other Options Are Wrong

(A) This incorrect answer does not consider the amount of water in the product when calculating the heat transferred.

(B) When calculating the enthalpy change of the combustion reaction, this incorrect answer omits the amount of n-pentane that reacted.

(C) When calculating the enthalpy change of the combustion reaction, this answer determines an incorrect value for the enthalpy of the CO formed.

SOLUTION 19

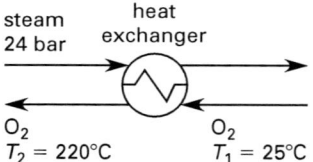

The molar flow rate of oxygen in the heat exchanger, F, is given as 150 kmol/h. The average heat capacity of oxygen, $\overline{C}_p$, is given as 30.07 kJ/kmol·°C. Because the heat exchanger is insulated, it operates adiabatically.

From an energy balance around the heat exchanger, the saturated steam provides the heat that the oxygen stream requires to go from 25°C to 220°C. The required enthalpy difference of the oxygen stream is

$$\Delta H = \overline{C}_p(T_2 - T_1)$$
$$= \left(30.07 \, \frac{\text{kJ}}{\text{kmol·°C}}\right)(220°C - 25°C)$$
$$= 5863.65 \text{ kJ/kmol}$$

The condensation enthalpy of saturated steam at a pressure of 24 bar, λ, from the steam tables, is 1858.14 kJ/kg. The steam mass flow rate is $\dot{m}_S$. The energy balance around the heat exchanger is

$$\dot{m}_S \lambda = F \Delta H$$

Solving for the steam mass flow rate gives

$$\dot{m}_S = \frac{F \Delta H}{\lambda} = \frac{\left(150 \, \frac{\text{kmol}}{\text{h}}\right)\left(5863.65 \, \frac{\text{kJ}}{\text{kmol}}\right)}{1858.14 \, \frac{\text{kJ}}{\text{kg}}}$$
$$= 475.1 \text{ kg/h} \quad (480 \text{ kg/h})$$

The answer is (D).

Why Other Options Are Wrong

(A) This incorrect answer fails to calculate the enthalpy difference required by the oxygen stream. This answer simply divides the molar flow rate of oxygen by the condensation enthalpy of the saturated steam.

(B) This incorrect answer calculates the saturated steam mass rate by taking a feed rate of oxygen equal to 1 kmol/h instead of 150 kmol/h.

(D) This incorrect answer uses the condensation enthalpy of the saturated steam at a pressure of 1.5 bar (the pressure of the oxygen stream) instead of 2 bar (the pressure of the saturated steam).

SOLUTION 20

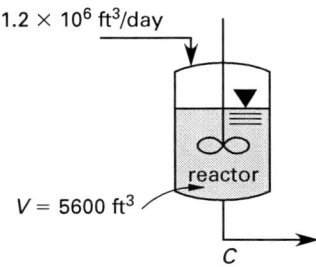

The tank volume, V, is given as 5600 ft^3. The volumetric flow rate through the reactor is

$$Q = \left(1.2 \times 10^6 \ \frac{\text{ft}^3}{\text{day}}\right)\left(\frac{1 \ \text{day}}{24 \ \text{hr}}\right)$$
$$= 50{,}000 \ \text{ft}^3/\text{hr}$$

The hydraulic residence time in the tank is

$$\theta = \frac{V}{Q} = \frac{5600 \ \text{ft}^3}{50{,}000 \ \frac{\text{ft}^3}{\text{hr}}}$$
$$= 0.112 \ \text{hr}$$

The input concentration of S, C_{in}, is given as 0.0125 lbmol/ft^3. The first-order decay rate, k, is given as 0.25 hr^{-1}. Because the reaction takes place in a CSTR, the exit concentration of S from the reactor is identical to the concentration of S inside the reactor. For a first-order decay, the reactor concentration of S is

$$C = \frac{C_{\text{in}}}{1+k\theta} = \frac{0.0125 \ \frac{\text{lbmol}}{\text{ft}^3}}{1+\left(\dfrac{0.25}{\text{hr}}\right)(0.112 \ \text{hr})}$$
$$= 0.0122 \ \text{lbmol/ft}^3 \quad (0.012 \ \text{lbmol/ft}^3)$$

The answer is (B).

Why Other Options Are Wrong

(A) This answer uses an incorrect expression for the first-order decay. This answer uses $C = C_{\text{in}}e^{-(1+k\theta)}$ instead of $C = C_{\text{in}}/(1+k\theta)$.

(C) This answer erroneously assumes S will reach a concentration that is the average of the steady-state concentration and the new concentration.

(D) This incorrect answer is the new input concentration of S. This option fails to recognize that the question refers to the steady-state concentration of the system before the input concentration changed. Therefore, the new value of the concentration of S has no effect on the correct answer.

SOLUTION 21

The feed mass flow rate, $\dot{m}_1$, is given as 12,000 lbm/hr. The total mass balance around the splitter gives

$$\dot{m}_1 = \dot{m}_2 + \dot{m}_3 = 12{,}000 \ \text{lbm/hr}$$

The mass fraction of suspended solids in the feed, x_1, is given as 0.10.

Because the splitter does not change the fraction of suspended solids in the streams, the mass fraction of suspended solids in stream 2, x_2, equals the mass fraction of suspended solids in stream 3, x_3.

$$x_1 = x_2 = x_3$$
$$= 0.10$$

The mass flow rate of $\dot{m}_3$ is given as 12% of the feed or

$$\dot{m}_3 = 0.12\dot{m}_1 = (0.12)\left(12{,}000 \ \frac{\text{lbm}}{\text{hr}}\right)$$
$$= 1440 \ \text{lbm/hr}$$

A suspended solids mass balance around the splitter gives

$$x_1\dot{m}_1 = x_2\dot{m}_2 + x_3\dot{m}_3$$
$$\dot{m}_2 = \frac{x_1\dot{m}_1 - x_3\dot{m}_3}{x_2}$$
$$= \frac{(0.10)\left(12{,}000 \ \dfrac{\text{lbm}}{\text{hr}}\right) - (0.10)\left(1440 \ \dfrac{\text{lbm}}{\text{hr}}\right)}{0.10}$$
$$= 10{,}560 \ \text{lbm/hr}$$

The mass fraction of suspended solids in the outlet from the evaporator, x_5, is given as 0.85. A total mass balance around the evaporator gives

$$\dot{m}_2 = \dot{m}_4 + \dot{m}_5 = 10{,}560 \ \text{lbm/hr}$$

The water mass balance around the evaporator gives

$$(1-x_2)\dot{m}_2 = \dot{m}_4 + (1-x_5)\dot{m}_5$$
$$\dot{m}_5 = \frac{\dot{m}_2 - (1-x_2)\dot{m}_2}{x_5}$$
$$= \frac{10{,}560 \ \dfrac{\text{lbm}}{\text{hr}} - (1-0.10)\left(10{,}560 \ \dfrac{\text{lbm}}{\text{hr}}\right)}{0.85}$$
$$= 1242 \ \text{lbm/hr}$$

Solving the total mass balance and the water mass balance equations around the evaporator gives

$$\dot{m}_4 = \dot{m}_2 - \dot{m}_5 = 10{,}560 \ \frac{\text{lbm}}{\text{hr}} - 1242 \ \frac{\text{lbm}}{\text{hr}}$$
$$= 9318 \ \text{lbm/hr}$$

The total mass balance around the mixer gives

$$\dot{m}_6 = \dot{m}_5 + \dot{m}_3$$
$$= 1242 \ \frac{\text{lbm}}{\text{hr}} + 1440 \ \frac{\text{lbm}}{\text{hr}}$$
$$= 2682 \ \text{lbm/hr}$$

The suspended solids balance around the mixer gives

$$x_6 \dot{m}_6 = x_5 \dot{m}_5 + x_3 \dot{m}_3$$
$$x_6 = \frac{x_5 \dot{m}_5 + x_3 \dot{m}_3}{\dot{m}_6}$$
$$= \frac{(0.85)\left(1242 \ \frac{\text{lbm}}{\text{hr}}\right) + (0.10)\left(1440 \ \frac{\text{lbm}}{\text{hr}}\right)}{2682 \ \frac{\text{lbm}}{\text{hr}}}$$
$$= 0.45$$

The answer is (B).

Why Other Options Are Wrong

(A) This answer erroneously calculates the suspended solids balance around the mixer by adding the mass flow rates of streams 6 and 3 instead of adding streams 3 and 5.

(C) This incorrect answer takes the average of the mass flow rates of the suspended solids of the two streams to the mixer or the average of the suspended solids of stream 5 and stream 3. This answer assumes that the resulting stream 6 contains an average of the mass flow rates of the suspended solids of the streams comprising it.

(D) This incorrect answer adds the mass flow rates of the suspended solids of the two streams to the mixer, assuming the mass flow rates of the suspended solids is an additive property. This answer assumes the mass flow rates of the suspended solids of the resulting stream 6 equals the mass flow rates of the suspended solids of the mass flow rates of the component streams, streams 5 and 3.

SOLUTION 22

The molar flow rate of the feed, F_F, is given as 1000 lbmol/hr. The molar flow rate in stream 4 is F_4. Because the molar flow rate in stream 3 is one third of the molar flow rate in stream 4, the molar flow rate in stream 3 is

$$F_3 = \tfrac{1}{3} F_4$$

Because the molar flow rate of the liquid in stream 2 is twice the molar flow rate of the liquid in stream 3, the molar flow rate of the liquid in stream 2 is

$$F_2 = 2 F_3$$

Substituting,

$$F_2 = (2)\tfrac{1}{3} F_4 = \tfrac{2}{3} F_4$$

A mass balance around splitter S1 gives

$$F_1 = F_2 + F_3 + F_4$$

Expressing F_2 and F_3 in terms of F_4 gives

$$1000 \ \frac{\text{lbmol}}{\text{hr}} = \tfrac{2}{3} F_4 + \tfrac{1}{3} F_4 + F_4 = 2 F_4$$

$$F_4 = \frac{F_1}{2} = \frac{1000 \ \frac{\text{lbmol}}{\text{hr}}}{2}$$
$$= 500 \ \text{lbmol/hr}$$

Because the molar flow rate of the liquid in stream 3 is one third of the molar flow rate of the liquid in stream 4, the molar flow rate in stream 3 is

$$F_3 = \left(\frac{1}{3}\right)\left(500 \ \frac{\text{lbmol}}{\text{hr}}\right) = 500/3 \ \text{lbmol/hr}$$

Because stream 3 is divided in two streams of equal molar flow rates, the molar flow rate of the liquid in stream 6 is

$$F_6 = \tfrac{1}{2} F_3 = \left(\frac{1}{2}\right)\left(\frac{500 \ \frac{\text{lbmol}}{\text{hr}}}{3}\right)$$
$$= \left(\frac{1}{6}\right)(500 \ \text{lbmol/hr})$$

Because the liquid in stream 6 combines with the liquid in stream 4, a mole balance around mixer 1 gives the molar flow rate of stream 7. The molar flow rate of the liquid in stream 7 is

$$F_7 = F_4 + F_6 = 500 \ \frac{\text{lbmol}}{\text{hr}} + \left(\frac{1}{6}\right)\left(500 \ \frac{\text{lbmol}}{\text{hr}}\right)$$
$$= 583.33 \ \text{lbmol/hr} \quad (580 \ \text{lbmol/hr})$$

The answer is (C).

Why Other Options Are Wrong

(A) This incorrect answer adds F_2 and F_6 instead of adding F_4 and F_6 when calculating F_7.

(B) This incorrect answer calculates F_6 as one third of F_3 instead of one half of F_3.

(D) This incorrect answer calculates F_3 as one half of F_4 instead of one third of F_4.

SOLUTION 23

Performing a mass balance around the mixer gives the mass flow rate of the hydrocarbon stream to the reactor, $\dot{m}_{\text{HC}}$.

The mass flow rate of the hydrocarbon stream to the reactor is the total mass of hydrocarbons flowing into the reactor.

$$\begin{aligned} \dot{m}_{\text{HC}} &= F_{\text{C}_4\text{H}_8}\text{MW}_{\text{C}_4\text{H}_8} + F_{\text{C}_4\text{H}_{10}}\text{MW}_{\text{C}_4\text{H}_{10}} \\ &\quad + F_{\text{iC}_4\text{H}_{10}}\text{MW}_{\text{iC}_4\text{H}_{10}} + F_{\text{C}_8\text{H}_{18}}\text{MW}_{\text{C}_8\text{H}_{18}} \\ &= \left(217\ \frac{\text{kmol}}{\text{h}}\right)\left(56.10\ \frac{\text{kg}}{\text{kmol}}\right) \\ &\quad + \left(17\,270\ \frac{\text{kmol}}{\text{h}}\right)\left(58.12\ \frac{\text{kg}}{\text{kmol}}\right) \\ &\quad + \left(43\,391\ \frac{\text{kmol}}{\text{h}}\right)\left(58.12\ \frac{\text{kg}}{\text{kmol}}\right) \\ &\quad + \left(8418\ \frac{\text{kmol}}{\text{h}}\right)\left(114.22\ \frac{\text{kg}}{\text{kmol}}\right) \\ &= 4\,499\,295\ \text{kg/h} \end{aligned}$$

The molar flow rate of iC_4H_{10} reacted, $F_{\text{iC}_4\text{H}_{10},\text{reacted}}$, is given as 191 kmol/h. A mole balance of iC_4H_{10} around the reactor gives the molar flow rate of iC_4H_{10} out of the reactor.

$$\begin{aligned} F_{\text{iC}_4\text{H}_{10},\text{RO}} &= F_{\text{iC}_4\text{H}_{10}} - F_{\text{iC}_4\text{H}_{10},\text{reacted}} \\ &= 43\,391\ \frac{\text{kmol}}{\text{h}} - 191\ \frac{\text{kmol}}{\text{h}} \\ &= 43\,200\ \text{kmol/h} \end{aligned}$$

The molar flow rate of iC_4H_{10} to the decanter, $F_{\text{iC}_4\text{H}_{10},\text{decanter}}$, is given as 1300 kmol/h. Because the distillation column bottoms contain pure iC_4H_{10}, a molar balance of iC_4H_{10} around the distillation column gives

$$F_{\text{iC}_4\text{H}_{10},\text{bottom}} = F_{\text{iC}_4\text{H}_{10},\text{decanter}} = 1300\ \text{kmol/h}$$

A mole balance of iC_4H_{10} around the decanter gives the molar flow rate of iC_4H_{10} in the emulsion.

$$\begin{aligned} F_{\text{iC}_4\text{H}_{10},\text{emul}} &= F_{\text{iC}_4\text{H}_{10},\text{RO}} - F_{\text{iC}_4\text{H}_{10},\text{bottom}} \\ &= 43\,200\ \frac{\text{kmol}}{\text{h}} - 1300\ \frac{\text{kmol}}{\text{h}} \\ &= 41\,900\ \text{kmol/h} \end{aligned}$$

The splitter ratio is the molar ratio of the molar flow rate of iC_4H_{10} in the emulsion stream to the molar flow rate of iC_4H_{10} out of the reactor. The splitter ratio is

$$\begin{aligned} E &= \frac{F_{\text{iC}_4\text{H}_{10},\text{emul}}}{F_{\text{iC}_4\text{H}_{10},\text{RO}}} = \frac{41\,900\ \frac{\text{kmol}}{\text{h}}}{43\,200\ \frac{\text{kmol}}{\text{h}}} \\ &= 0.9699 \end{aligned}$$

Because the ratio of the mass flow rate of H_2SO_4 to the mass flow rate of hydrocarbon in the bottom stream out of the decanter is given as 2, the mass flow rate of the acid stream to the reactor is

$$\begin{aligned} \dot{m}_{\text{H}_2\text{SO}_4} &= 2\dot{m}_{\text{HC}} = (2)\left(4\,499\,295\ \frac{\text{kg}}{\text{h}}\right) \\ &= 8\,998\,590\ \text{kg/h} \end{aligned}$$

The ratio of the mass flow rate of H_2SO_4 in the emulsion stream to the mass flow rate of H_2SO_4 out of the reactor equals E. The mass flow rate of H_2SO_4 in the emulsion is

$$\begin{aligned} \dot{m}_{\text{H}_2\text{SO}_4,\text{emul}} &= \dot{m}_{\text{H}_2\text{SO}_4} E \\ &= \left(8\,998\,590\ \frac{\text{kg}}{\text{h}}\right)(0.9699) \\ &= 8\,727\,732\ \text{kg/h} \end{aligned}$$

A mass balance of acid around the mixer gives the mass flow rate of acid recycled to the reactor. The mass flow rate of acid in the recycle stream is

$$\begin{aligned} \dot{m}_{\text{H}_2\text{SO}_4,\text{recycled}} &= \dot{m}_{\text{H}_2\text{SO}_4} - \dot{m}_{\text{H}_2\text{SO}_4,\text{emul}} \\ &= 8\,998\,590\ \frac{\text{kg}}{\text{h}} - 8\,727\,732\ \frac{\text{kg}}{\text{h}} \\ &= 270\,858\ \text{kg/h} \quad (2.7 \times 10^5\ \text{kg/h}) \end{aligned}$$

The answer is (B).

Why Other Options Are Wrong

(A) This incorrect answer omits the molar flow rate of C_8H_{18} in the feed.

(C) This incorrect answer is the mass flow rate of the hydrocarbon stream to the reactor.

(D) This incorrect answer excludes the splitter ratio. This answer is calculated with 100% of the H_2SO_4 in the recycle stream and none in the emulsion. In this answer, the decanter recovery factor, E, is 1.

SOLUTION 24

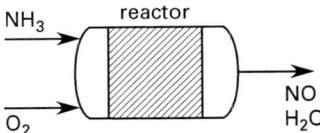

The conversion of the limiting reactant, X, is given as 0.85. The molar flow rate of the feed, F_F, is given as 120 lbmol/hr. Because the feed is an equimolar mixture and consists of two components, NH_3 and O_2, the input rate of NH_3 is

$$F_{in,NH_3} = \frac{F_F}{2} = \frac{120\ \frac{lbmol}{hr}}{2}$$
$$= 60\ lbmol/hr$$

Similarly,

$$F_{in,O_2} = \frac{F_F}{2} = \frac{120\ \frac{lbmol}{hr}}{2}$$
$$= 60\ lbmol/hr$$

Because the feed does not contain NO, the input rate of NO, $F_{in,NO}$, is 0 lbmol/hr. Similarly, the input rate of water, F_{in,H_2O}, is 0 lbmol/hr. From the balanced reaction, the stoichiometric coefficients are

component	symbol	coefficient
NH_3	σ_{NH_3}	4
O_2	σ_{O_2}	5
NO	σ_{NO}	4
H_2O	σ_{H_2O}	6

For NH_3, the ratio of the molar flow rate in the feed to the stoichiometric coefficient is

$$F_{NH_3} = \frac{F_{in,NH_3}}{\sigma_{NH_3}} = \frac{60\ \frac{lbmol}{hr}}{4}$$
$$= 15\ lbmol/hr$$

For O_2, the ratio of the molar flow rate in the feed to the stoichiometric coefficient is

$$F_{O_2} = \frac{F_{in,O_2}}{\sigma_{O_2}} = \frac{60\ \frac{lbmol}{hr}}{5}$$
$$= 12\ lbmol/hr$$

Because F_{NH_3} is larger than F_{O_2}, oxygen is the limiting reactant. The reactant that is first depleted is oxygen. For the entire reaction, the rate is

$$r = \frac{XF_{in,O_2}}{\sigma_{O_2}} = \frac{(0.85)\left(60\ \frac{lbmol}{hr}\right)}{5}$$
$$= 10.20\ lbmol/hr$$

A mole balance of NH_3 around the reactor gives the output rate for NH_3.

$$F_{out,NH_3} = F_{in,NH_3} - \sigma_{NH_3}r$$
$$= 60\ \frac{lbmol}{hr} - (4)\left(10.20\ \frac{lbmol}{hr}\right)$$
$$= 19.20\ lbmol/hr$$

A mole balance of O_2 around the reactor gives the output rate for O_2.

$$F_{out,O_2} = F_{in,O_2} - \sigma_{O_2}r$$
$$= 60\ \frac{lbmol}{hr} - (5)\left(10.20\ \frac{lbmol}{hr}\right)$$
$$= 9.00\ lbmol/hr$$

A mole balance of NO around the reactor gives the output rate for NO.

$$F_{out,NO} = F_{in,NO} + \sigma_{NO}r$$
$$= 0\ \frac{lbmol}{hr} + (4)\left(10.20\ \frac{lbmol}{hr}\right)$$
$$= 40.80\ lbmol/hr$$

A mole balance of H_2O around the reactor gives the output rate for H_2O.

$$F_{out,H_2O} = F_{in,H_2O} + \sigma_{H_2O}r$$
$$= 0\ \frac{lbmol}{hr} + (6)\left(10.20\ \frac{lbmol}{hr}\right)$$
$$= 61.20\ lbmol/hr$$

The mole balance around the reactor gives the total output rate of the product stream.

$$F_p = F_{out,NH_3} + F_{out,O_2} + F_{out,NO} + F_{out,H_2O}$$
$$= 19.20\ \frac{lbmol}{hr} + 9.00\ \frac{lbmol}{hr}$$
$$+ 40.80\ \frac{lbmol}{hr} + 61.20\ \frac{lbmol}{hr}$$
$$= 130.20\ lbmol/hr\quad (130\ lbmol/hr)$$

The answer is (B).

Why Other Options Are Wrong

(A) This answer erroneously subtracts the value of the molar flow rate of water instead of adding it when calculating the total molar flow rate of production.

(C) This incorrect answer does not calculate the value of the molar flow rate for the entire reaction. This answer uses F_{O_2} instead of r when calculating the output from the reactor of each component.

(D) This incorrect answer does not calculate the value of the molar flow rate for the entire reaction. This answer uses F_{NH_3} instead of r when calculating the output from the reactor of each component.

SOLUTION 25

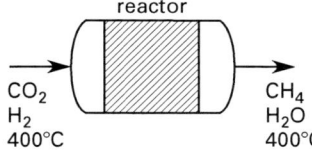

The universal gas constant, R, is 8.314 J/mol·K. The thermodynamic data is given at

$$T_1 = 25°C + 273° = 298K$$

The temperature at which the reaction takes place is given as

$$T_2 = 400°C + 273° = 673K$$

From the balanced reaction given in the problem statement, the stoichiometric coefficients are as follows.

species	coefficient, σ
CO_2	1
H_2	4
CH_4	1
H_2O	2

Because the products are CH_4 and H_2O, the enthalpy change for the products, at 25°C, is

$$\Delta H^0_{products} = \sigma_{CH_4}\Delta H^0_{f,CH_4} + \sigma_{H_2O}\Delta H^0_{f,H_2O}$$
$$= (1)\left(-74.87\ \frac{kJ}{mol}\right) + (2)\left(-241.83\ \frac{kJ}{mol}\right)$$
$$= -558.53\ kJ/mol$$

At 25°C, the enthalpy change for the reaction is

$$\Delta H^0_{298} = \Delta H^0_{products} - \Delta H^0_{reactants}$$
$$= -558.53\ \frac{kJ}{mol} - \left(-393.51\ \frac{kJ}{mol}\right)$$
$$= -165.02\ kJ/mol$$

At 25°C, the change in the Gibbs free-energy of formation for the reactants is

$$\Delta G_{reactants} = \sigma_{CO_2}\Delta G^0_{f,CO_2} + \sigma_{H_2}\Delta G^0_{f,H_2}$$
$$= (1)\left(-394,359\ \frac{J}{mol}\right) + (4)\left(0\ \frac{J}{mol}\right)$$
$$= -394,359\ J/mol$$

At 25°C, the change in the Gibbs free-energy of formation for the products is

$$\Delta G_{products} = \sigma_{CH_4}\Delta G^0_{f,CH_4} + \sigma_{H_2O}\Delta G^0_{f,H_2O}$$
$$= (1)\left(-50,460\ \frac{J}{mol}\right)$$
$$+ (2)\left(-228,572\ \frac{J}{mol}\right)$$
$$= -507,604\ J/mol$$

At 25°C, the Gibbs free-energy change for the reaction is

$$\Delta G^0_{298} = \Delta G_{products} - \Delta G_{reactants}$$
$$= -507,604\ \frac{J}{mol} - \left(-394,359\ \frac{J}{mol}\right)$$
$$= -113,245\ J/mol \quad (-113.245\ kJ/mol)$$

The average heat capacity of the products is

$$\overline{C}_{p,products} = \sigma_{CH_4}\overline{C}_{p,CH_4} + \sigma_{H_2O}\overline{C}_{p,H_2O}$$
$$= (1)\left(50.81\ \frac{J}{mol\cdot K}\right) + (2)\left(34.71\ \frac{J}{mol\cdot K}\right)$$
$$= 120.23\ J/mol\cdot K$$

The heat capacity of the reactants is

$$\overline{C}_{p,reactants} = \sigma_{CO_2}\overline{C}_{p,CO_2} + \sigma_{H_2}4\overline{C}_{p,H_2}$$
$$= (1)\left(52.55\ \frac{J}{mol\cdot K}\right) + (4)\left(30.12\ \frac{J}{mol\cdot K}\right)$$
$$= 173.03\ J/mol\cdot K$$

The heat capacity of the reaction is

$$\Delta\overline{C}_p = \overline{C}_{p,products} - \overline{C}_{p,reactants}$$
$$= 120.23\ \frac{J}{mol\cdot K} - 173.03\ \frac{J}{mol\cdot K}$$
$$= -52.80\ J/mol\cdot K \quad (-0.05280\ kJ/mol\cdot K)$$

The Gibbs energy is

$$\Delta G = \Delta H - T\Delta S$$

Replacing and solving gives

$$\Delta S^0_{298} = \frac{\Delta H^0_{298} - \Delta G^0_{298}}{T}$$
$$= \frac{-165.02\ \frac{kJ}{mol} - \left(-113.245\ \frac{kJ}{mol}\right)}{298K}$$
$$= -0.1737\ kJ/mol\cdot K$$

The entropy change of the reaction at 400°C is

$$\Delta S_{673} = \Delta S^0_{298} + \Delta C_p \int_{298K}^{673K} \frac{dT}{T}$$
$$= -0.1737\ \frac{kJ}{mol\cdot K} + \left(-0.05280\ \frac{kJ}{mol\cdot K}\right)$$
$$\times \ln\frac{673K}{298K}$$
$$= -0.216714\ kJ/mol\cdot K$$

At 400°C, the enthalpy change for the reaction is

$$\Delta H_{673} = \Delta H^0_{298} + \Delta C_p \int_{298K}^{673K} dT$$

$$= -165.02 \, \frac{\text{kJ}}{\text{mol}}$$

$$+ \left(-0.05280 \, \frac{\text{kJ}}{\text{mol·K}}\right)(673K - 298K)$$

$$= -184.8200 \, \text{kJ/mol}$$

At 673K, the Gibbs free-energy for the reaction is

$$\Delta G_{673} = \Delta H_{673} - T\Delta S_{673}$$

$$= -184.8200 \, \frac{\text{kJ}}{\text{mol}}$$

$$- (673K)\left(-0.216714 \, \frac{\text{kJ}}{\text{mol·K}}\right)$$

$$= -38.9715 \, \text{kJ/mol}$$

At 673K, the reaction constant is

$$\ln K_{\text{rxn},673} = -\frac{\Delta G^0_T}{RT} = -\frac{-38{,}971.5 \, \frac{\text{J}}{\text{mol}}}{\left(8.314 \, \frac{\text{J}}{\text{mol·K}}\right)(673K)}$$

$$= 6.9650$$

$$K_{\text{rxn},673} = 1058.91$$

Let the reference pressure, p^0, be 1 atm. The total pressure, p, is 1 atm.

From the stoichiometry of the reaction

component	initial (mol)	reacted	products	mole fraction, y
CO_2	1.0	$-x$	$1-x$	$\frac{1-x}{5-2x}$
H_2	4.0	$-4x$	$4-4x$	$\frac{4-4x}{5-2x}$
CH_4	0	x	x	$\frac{x}{5-2x}$
H_2O	0	$2x$	$2x$	$\frac{2x}{5-2x}$
Total	5.0	$-2x$	$5-2x$	1

The concentration equilibrium constant is

$$K_{\text{rxn}} = \frac{y_{CH_4}\left(\frac{p}{p^0}\right)\left(y_{H_2O}\left(\frac{p}{p^0}\right)\right)^2}{y_{CO_2}\left(\frac{p}{p^0}\right)\left(y_{H_2}\left(\frac{p}{p^0}\right)\right)^4}$$

$$= \frac{y_{CH_4}(y_{H_2O})^2}{y_{CO_2}(y_{H_2})^4 \left(\frac{p}{p^0}\right)^2}$$

$$= 1058.93 \, \text{L}^2/\text{mol}^2$$

$$K = \frac{\left(\frac{x}{5-2x}\right)\left(\frac{2x}{5-2x}\right)^2}{\left(\frac{1-x}{5-2x}\right)\left(\frac{4-4x}{5-2x}\right)^4 \left(\frac{1 \, \text{atm}}{1 \, \text{atm}}\right)^2}$$

$$= \frac{x^3(2x-5)^2}{64(1-x)^5}$$

$$= 1058.93 \, \text{L}^2/\text{mol}^2$$

Solving for x using a programmable calculator gives

$$x = 0.842485 \, \text{mol} \quad (0.84 \, \text{mol})$$

The answer is (B).

Alternate Solution

The preceding equation could also be solved iteratively by giving values to x. The value of x that makes K equal to 1058.93 L^2/mol^2 is the solution.

Solving for x,

x	x^3	$(2x-5)^2$	$64(1-x)^5$	K
0.90	0.729	10.24	0.00064	11.664
0.85	0.614	10.89	0.00486	1376.0949
0.84	0.593	11.0224	0.006744	973.49593
0.842485	0.598	10.98942	0.006206	1058.9413

Why Other Options Are Wrong

(A) This incorrect answer omits the exponent 2 in the numerator of the equilibrium constant.

(C) This incorrect answer omits the exponent 4 in the denominator of the equilibrium constant.

(D) This answer uses an incorrect value for the universal gas constant when calculating the concentration equilibrium constant.

SOLUTION 26

The mass flow rate, $\dot{m}$, is given as 1 kg/s. Represent the devices by the following pairs of states: pump 1-2, boiler 2-3, HP turbine 3-4, LP reboiler 4-5, LP turbine 5-6, condenser 6-1.

For State 1:

At state 1 the working fluid is saturated water in the liquid phase at its saturation temperature. The pressure, p_1, is given as 12.0 kPa. The mass fraction of the vapor in the working fluid at state 1, x_1, is 0.0. From the steam tables, the entropy of the working fluid at state 1, s_1, is 0.696 kJ/kg·K. From the steam tables, the enthalpy of the fluid at state 1, h_1, is 206.839 kJ/kg.

For State 6:

At state 6 the working fluid is a mixture of liquid and vapor. The pressure at state 6, p_6, is 12.0 kPa. The moisture content in the low-pressure turbine is not to exceed 8%. The mass fraction of the vapor, x_6, is 0.08. From the steam tables, the temperature at state 6, T_6, is 49.404°C. From the steam tables, the entropy of the fluid at state 6 is

$$s_6 = 7.496 \ \frac{\text{kJ}}{\text{kg·K}}$$
$$= s_1 + (1 - x_6)\Delta s$$

From the steam tables, the latent heat of water at 49.404°C, λ, is 2383.704 kJ/kg. The enthalpy of the fluid at state 6 is

$$h_6 = h_1 + (1 - x_6)\lambda$$
$$= 206.839 \ \frac{\text{kJ}}{\text{kg}}$$
$$+ (1 - 0.08)\left(2383.704 \ \frac{\text{kJ}}{\text{kg}}\right)$$
$$= 2399.847 \ \text{kJ/kg}$$

For State 5:

At state 5 the working fluid is a superheated vapor. The working fluid is above its saturation temperature, so there is no condensation. The entropy of the fluid at state 5 is the same as the entropy of the fluid at state 6 because the work done by the fluid in the low-pressure turbine is reversible and adiabatic (isentropic). The entropy at state 5, s_5, is 7.496 kJ/kg·K. Because the temperature of the fluid at state 5 is the same as the temperature at state 3, $T_5 = 612°C$.

From the steam tables and with the values of entropy and temperature at state 5, the enthalpy at state 5, h_5, is 3707.727 kJ/kg. From the steam tables the pressure at state 5, p_5, is 3277.419 kPa.

For State 2:

At state 2 the working fluid is subcooled liquid water. The working fluid is below its saturation temperature, so there is no vapor. The pressure at state 2, p_2, is given as 12 MPa. The entropy at state 2 is the same as the entropy at state 1 because the work done on the fluid in the pump is reversible and adiabatic (isentropic). The entropy at state 2, s_2, is 0.696 kJ/kg·K. From the steam tables the enthalpy at state 2, h_2, is 218.938 kJ/kg.

For State 3:

At state 3 the working fluid is a superheated vapor. The working fluid is above its saturation temperature, so there is no condensation. The pressure at state 3 is the same as the pressure at state 2 because the reboiler does not change the pressure of the working fluid. The pressure at state 3, p_3, is 12 MPa. The temperature of the fluid at state 3, T_3, is given as 612°C. From the steam tables and with the value of the pressure at state 3 and the temperature at state 3, the entropy at state 3, s_3, is 6.837 kJ/kg·K. From the steam tables, the enthalpy at state 3, h_3, is 3638.450 kJ/kg.

For State 4:

At state 4 the working fluid is a superheated vapor. The working fluid is above its saturation temperature, so there is no condensation. The pressure at state 4 is the same as the pressure at state 5 because the reboiler does not change the pressure of the working fluid. The pressure at state 4, p_4, equals the pressure at state 5, p_5.

$$p_4 = p_5 = 3277.419 \ \text{kPa}$$

The entropy of the working fluid at state 4 is the same as the entropy at state 3 because the work done by the fluid in the high-pressure turbine is reversible and adiabatic (isentropic). The entropy at state 4, s_4, equals the entropy at state 3, s_3.

$$s_4 = s_3 = 6.837 \ \text{kJ/kg·K}$$

From the steam tables the enthalpy of the fluid at state 4, h_4, is 3201.524 kJ/kg.

state	pressure (kPa)	temperature (°C)	enthalpy (kJ/kg)	entropy (kJ/kg·K)	phase
1	12.0	49.404	206.839	0.696	saturated water
2	12,000	49.404	218.938	0.696	subcooled liquid
3	12,000	612	3638.450	6.837	superheated vapor
4	3227.419	389.530	3201.524	6.837	superheated vapor
5	3227.419	612	3707.727	7.496	superheated vapor
6	12.0	49.404	2399.846	7.496	mixture

Using the convention that the work done by the system on the surroundings is positive, the work done by the pump on 1 kg/s of working fluid when the working fluid goes from state 1 to state 2 is

$$W_{\text{pump}} = \dot{m}(h_1 - h_2)$$
$$= \left(1 \ \frac{\text{kg}}{\text{s}}\right)\left(206.839 \ \frac{\text{kJ}}{\text{kg}} - 218.938 \ \frac{\text{kJ}}{\text{kg}}\right)$$
$$= -12.100 \ \text{kJ/s} \quad (-12.100 \ \text{kW})$$

Using the convention that the heat going from the surroundings to the system is positive, the heat transferred by the boiler to the working fluid as the working fluid goes from state 2 to state 3 is

$$Q_{\text{boiler}} = \dot{m}(h_3 - h_2)$$
$$= \left(1 \ \frac{\text{kg}}{\text{s}}\right)\left(3638.450 \ \frac{\text{kJ}}{\text{kg}} - 218.938 \ \frac{\text{kJ}}{\text{kg}}\right)$$
$$= 3419.512 \ \text{kJ/s} \quad (3419.512 \ \text{kW})$$

The work done by the working fluid in the high-pressure turbine as the working fluid goes from state 3 to state 4 is

$$W_{\text{HP}} = \dot{m}(h_3 - h_4)$$
$$= \left(1 \; \frac{\text{kg}}{\text{s}}\right)\left(3638.450 \; \frac{\text{kJ}}{\text{kg}} - 3201.524 \; \frac{\text{kJ}}{\text{kg}}\right)$$
$$= 436.926 \; \text{kJ/s} \quad (436.926 \; \text{kW})$$

The heat transferred by the reboiler to the working fluid as the working fluid goes from state 4 to state 5 is

$$Q_{\text{reboiler}} = \dot{m}(h_5 - h_4)$$
$$= \left(1 \; \frac{\text{kg}}{\text{s}}\right)\left(3707.727 \; \frac{\text{kJ}}{\text{kg}} - 3201.524 \; \frac{\text{kJ}}{\text{kg}}\right)$$
$$= 506.203 \; \text{kJ/s} \quad (506.203 \; \text{kW})$$

The work done by the working fluid in the low-pressure turbine as the working fluid goes from state 5 to state 6 is

$$W_{\text{LP}} = \dot{m}(h_5 - h_6)$$
$$= \left(1 \; \frac{\text{kg}}{\text{s}}\right)\left(3707.727 \; \frac{\text{kJ}}{\text{kg}} - 2399.847 \; \frac{\text{kJ}}{\text{kg}}\right)$$
$$= 1307.880 \; \text{kJ/s} \quad (1307.880 \; \text{kW})$$

The net work done by the working fluid is

$$W_{\text{net}} = W_{\text{pump}} + W_{\text{LP}} + W_{\text{HP}}$$
$$= -12.100 \; \text{kW} + 436.926 \; \text{kW} + 1307.880 \; \text{kW}$$
$$= 1732.706 \; \text{kW}$$

The heat transferred from the reboiler to the working fluid when the working fluid goes from state 2 to state 3 plus the heat transferred from the reboiler to the working fluid when the working fluid goes from state 4 to state 5 is

$$Q_{\text{net}} = Q_{\text{boiler}} + Q_{\text{reboiler}}$$
$$= 3419.512 \; \text{kW} + 506.203 \; \text{kW}$$
$$= 3925.715 \; \text{kW}$$

The thermal efficiency is

$$\eta = \frac{W_{\text{net}}}{Q_{\text{net}}} \times 100\% = \frac{1732.706 \; \text{kW}}{3925.715 \; \text{kW}} \times 100\%$$
$$= 44.137\% \quad (44\%)$$

The answer is (A).

Why Other Options Are Wrong

(B) This incorrect answer does not add the heat transferred in the reboiler when calculating the efficiency. This answer accounts for the heat transferred in the boiler only.

(C) This incorrect answer is the thermal efficiency of the corresponding Carnot cycle.

(D) This incorrect answer adds the heat involved in each step of the cycle. This means that this answer adds the heat transferred in the boiler, reboiler, and condenser instead of using only the heat transferred in the boiler and reboiler when calculating the efficiency.

SOLUTION 27

The steam flowing between the inlet and the outlet from the diffuser comprises the system.

The heat flow rate, q, is given as 300 Btu/lbm. The mass flow rate of the steam, $\dot{m}$, is given as 250 lbm/hr. The gravitational acceleration, g, is 32.17 ft/sec^2. The gravitational constant, g_c, is 32.17 lbm-ft/lbf-sec^2.

The steady flow energy balance for flow from point 1 to point 2, with heat from the surroundings to the system taken as positive and work done by the system on the surroundings taken as positive, is

$$\dot{m}\left(\Delta h + \frac{\Delta \text{v}^2}{2g_c J} + \left(\frac{g}{g_c J}\right)\Delta z\right) = Q - W$$

Replacing the definition of enthalpy difference, $\Delta h = h_2 - h_1$, and solving for h_2 gives

$$h_2 = \frac{Q - W}{\dot{m}} + h_1 - \frac{\Delta \text{v}^2}{2g_c J} - \left(\frac{g}{g_c J}\right)\Delta z \quad \text{[I]}$$

The heat added to the stream in the system is

$$Q = q\dot{m} = \left(300 \; \frac{\text{Btu}}{\text{lbm}}\right)\left(250 \; \frac{\text{lbm}}{\text{hr}}\right)$$
$$= 75{,}000 \; \text{Btu/hr}$$

The steam undergoes both an elevation change and a change in velocity. The work done by the back pressure turbine is given as

$$W = (60 \; \text{hp})\left(550 \; \frac{\text{ft-lbf}}{\text{sec-hp}}\right)\left(\frac{1 \; \text{Btu}}{778 \; \text{ft-lbf}}\right)\left(3600 \; \frac{\text{sec}}{\text{hr}}\right)$$
$$= 152{,}699 \; \text{Btu/hr}$$

From the steam tables, the specific enthalpy of saturated steam at 356°F, h_1, is 1193.6 Btu/lbm. The velocity of the saturated steam at the inlet, v_1, is given as

130 ft/sec. The velocity of the saturated steam at the outlet, v_2, is given as 1 ft/sec. The kinetic energy is

$$\frac{\Delta v^2}{2g_c J} = \frac{v_2^2 - v_1^2}{2g_c J}$$

$$= \frac{\left(1 \frac{\text{ft}}{\text{sec}}\right)^2 - \left(130 \frac{\text{ft}}{\text{sec}}\right)^2}{(2)\left(32.17 \frac{\text{lbm-ft}}{\text{lbf-sec}^2}\right)\left(778 \frac{\text{ft-lbf}}{\text{Btu}}\right)}$$

$$= -0.338 \text{ Btu/lbm}$$

The potential energy term arising from a 220 ft elevation difference is

$$\left(\frac{g}{g_c J}\right)\Delta z = \left(\frac{g}{g_c J}\right)(z_2 - z_1)$$

$$= \left(\frac{32.17 \frac{\text{ft}}{\text{sec}^2}}{\left(32.17 \frac{\text{ft-lbm}}{\text{lbf-sec}^2}\right)\left(778 \frac{\text{ft-lbf}}{\text{Btu}}\right)}\right)$$

$$\times (220 \text{ ft} - 0 \text{ ft})$$

$$= 0.283 \text{ Btu/lbm}$$

Replacing the value in Eq. I gives the enthalpy at point 2.

$$h_2 = \frac{Q - W}{\dot{m}} + h_1 - \frac{\Delta v^2}{2g_c J} - \left(\frac{g}{g_c J}\right)\Delta z$$

$$= \frac{75{,}000 \frac{\text{Btu}}{\text{hr}} - 152{,}699 \frac{\text{Btu}}{\text{hr}}}{250 \frac{\text{lbm}}{\text{hr}}} + 1193.6 \frac{\text{Btu}}{\text{lbm}}$$

$$- \left(-0.338 \frac{\text{Btu}}{\text{lbm}}\right) - 0.283 \frac{\text{Btu}}{\text{lbm}}$$

$$= 882.9 \text{ Btu/lbm}$$

From the steam tables, the specific enthalpy of the saturated steam at 15 lbf/in^2, h_g, is 1150.9 Btu/lbm. From the steam tables, the specific enthalpy of the liquid at 15 lbf/in^2, h_l, is 181.4 Btu/lbm. The energy balance at point 2 is

$$h_2 = h_g x + h_l (1-x)$$

Solving for x gives

$$x = \frac{h_2 - h_l}{h_g - h_l} = \frac{882.9 \frac{\text{Btu}}{\text{lbm}} - 181.4 \frac{\text{Btu}}{\text{lbm}}}{1150.9 \frac{\text{Btu}}{\text{lbm}} - 181.4 \frac{\text{Btu}}{\text{lbm}}}$$

$$= 0.7236 \quad (0.7)$$

The answer is (C).

Why Other Options Are Wrong

(A) This incorrect answer assumes that the stream at the outlet of the diffuser contains no vapor.

(B) This incorrect answer uses a negative value for the heat added to the system when calculating the enthalpy at the diffuser.

(D) This incorrect answer assumes that the stream at the outlet of the diffuser contains no liquid. In this answer, this stream is gaseous.

SOLUTION 28

The isentropic exponent for air, k, is given as 1.40. The compressor inlet temperature is given as

$$T_1 = 70°\text{F} + 460° = 530°\text{R}$$

The compressor efficiency, η_c, is given as 0.87. The compressor pressure ratio, r_c, is given as 5. Because the compressor is isentropic, the exit temperature is

$$T_{2,s} = T_1 r_c^{\frac{k-1}{k}} = (530°\text{R})\left(5^{\frac{1.40-1}{1.40}}\right)$$

$$= 839.42°\text{R}$$

Because of inefficiency in the compressor, the true exit temperature is

$$T_2 = T_1 + \frac{T_{2,s} - T_1}{\eta_c} = 530°\text{R} + \frac{839.42°\text{R} - 530°\text{R}}{0.87}$$

$$= 886°\text{R}$$

The air heat capacity, $c_{p,\text{air}}$, is given as 0.24 Btu/lbm-°R. The compressor work is

$$W_c = c_{p,\text{air}}(T_1 - T_2)$$

$$= \left(0.24 \frac{\text{Btu}}{\text{lbm-°R}}\right)(530°\text{R} - 886°\text{R})$$

$$= -85.44 \text{ Btu/lbm}$$

The combustion chamber pressure drop, Δp_1, is given as 0.06. The turbine pressure ratio is

$$r_t = r_c(1 - \Delta p_1) = (5)(1 - 0.06)$$

$$= 4.7$$

The turbine inlet temperature is given as

$$T_3 = 1500°\text{F} + 460°$$

$$= 1960°\text{R}$$

The hot-gas heat capacity, $c_{p,g}$, is given as 0.2744 Btu/lbm-°R. The combustion chamber heat addition is

$$Q = c_{p,g}(T_3 - T_2)$$
$$= \left(0.2744 \; \frac{\text{Btu}}{\text{lbm-°R}}\right)(1960°\text{R} - 886°\text{R})$$
$$= 294.7 \; \text{Btu/lbm}$$

The hot-gas isentropic exponent, k_g, is given as 1.33. The turbine isentropic exit temperature is

$$T_{4,s} = \frac{T_3}{r_t^{(k_g-1)/k_g}} = \frac{1960°\text{R}}{4.7^{(1.33-1)/1.33}}$$
$$= 1335°\text{R}$$

The turbine isentropic efficiency, η_t, is given as 0.90. The turbine true exit temperature is

$$T_4 = T_3 - \eta_t(T_3 - T_{4,s})$$
$$= 1960°\text{R} - (0.90)(1960°\text{R} - 1335°\text{R})$$
$$= 1398°\text{R}$$

The turbine work is

$$W_t = c_{p,g}(T_3 - T_4)$$
$$= \left(0.2744 \; \frac{\text{Btu}}{\text{lbm-°R}}\right)(1960°\text{R} - 1398°\text{R})$$
$$= 154.21 \; \text{Btu/lbm}$$

The net work is

$$W_n = W_t + W_c = 154.21 \; \frac{\text{Btu}}{\text{lbm}} + \left(-85.44 \; \frac{\text{Btu}}{\text{lbm}}\right)$$
$$= 68.77 \; \text{Btu/lbm}$$

The thermal efficiency is

$$\eta_{\text{thermal}} = \frac{W_n}{Q} = \frac{68.77 \; \frac{\text{Btu}}{\text{lbm}}}{294.7 \; \frac{\text{Btu}}{\text{lbm}}}$$
$$= 0.2334 \quad (0.23)$$

The answer is (B).

Why Other Options Are Wrong

(A) This incorrect answer uses the value of $T_{2,s}$ instead of T_2 when calculating the heat added in the combustion chamber.

(C) This incorrect answer uses the value of $T_{4,s}$ instead of T_4 when calculating the work done by the turbine.

(D) This incorrect answer uses the wrong sign on the compressor work when calculating the net work.

SOLUTION 29

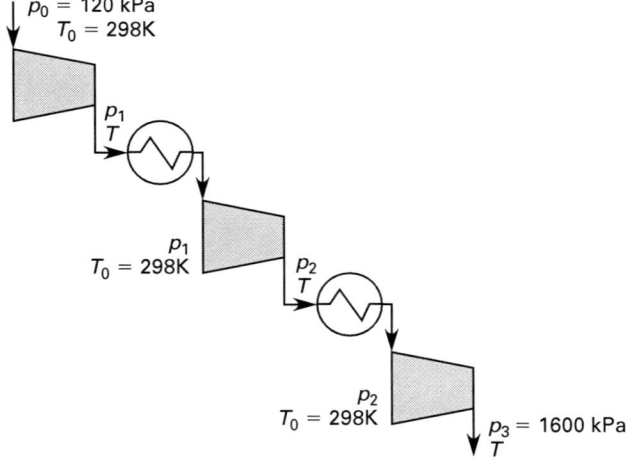

The molar flow rate of the ideal gas, F, is given as 15 mol/s. The initial temperature, T_0, is given as 298K. The initial pressure, p_0, is given as 120 kPa. The final pressure, p_3, is given as 1600 kPa. The number of stages, N, is given as 3. For an ideal gas, the ratio of the constant pressure heat capacity to the constant volume capacity, k, is 1.4. The efficiency of the compressor, η_c, is given as 0.9. The efficiency of the motor, η_m, is given as 0.7. The universal gas constant, R, is 8.314 J/mol·K.

For an ideal gas in isentropic compression with the intercoolers set to T_0 and equal compression ratios for each stage, the work done by the compressor in each stage is

$$W = \left(\frac{k}{k-1}\right)RT_0\left(\left(\frac{p}{p_0}\right)^{\frac{k-1}{k}} - 1\right)$$

Because the compression ratios are equal for each step,

$$\frac{p_1}{p_0} = \frac{p_2}{p_1} = \frac{p_3}{p_2}$$

For the first stage, the work done is

$$W_1 = \left(\frac{k}{k-1}\right)RT_0\left(\left(\frac{p_1}{p_0}\right)^{\frac{k-1}{k}} - 1\right)$$

For the second stage, the work done is

$$W_2 = \left(\frac{k}{k-1}\right)RT_0\left(\left(\frac{p_2}{p_1}\right)^{\frac{k-1}{k}} - 1\right)$$

For the third stage, the work done is

$$W_3 = \left(\frac{k}{k-1}\right) RT_0 \left(\left(\frac{p_3}{p_2}\right)^{\frac{k-1}{k}} - 1\right)$$

The total work is

$$\begin{aligned}W &= W_1 + W_2 + W_3 \\ &= \left(\frac{k}{k-1}\right) RT_0 \left(\left(\frac{p_1}{p_0}\right)^{\frac{k-1}{k}} - 1\right) \\ &\quad + \left(\frac{k}{k-1}\right) RT_0 \left(\left(\frac{p_2}{p_1}\right)^{\frac{k-1}{k}} - 1\right) \\ &\quad + \left(\frac{k}{k-1}\right) RT_0 \left(\left(\frac{p_3}{p_2}\right)^{\frac{k-1}{k}} - 1\right)\end{aligned}$$

Simplifying,

$$W = 3\left(\frac{k}{k-1}\right) RT_0 \left(\left(\frac{p_3}{p_0}\right)^{\frac{k-1}{3k}} - 1\right) \quad [\text{I}]$$

The power required by the compressor is

$$P = F\left(\frac{W}{\eta_m \eta_c}\right)$$

Replacing Eq. I in the preceding equation,

$$P = F\left(\frac{3\left(\frac{k}{k-1}\right) RT_0 \left(\left(\frac{p_3}{p_0}\right)^{\frac{k-1}{3k}} - 1\right)}{\eta_c \eta_m}\right)$$

$$= \left(15 \, \frac{\text{mol}}{\text{s}}\right)$$

$$\times \left(\frac{(3)\left(\frac{1.4}{1.4-1}\right)\left(8.314 \, \frac{\text{J}}{\text{mol·K}}\right)(298\text{K})}{(0.9)(0.7)}\right)$$

$$\times \left(\frac{1 \text{ kW}}{1000 \text{ J/s}}\right)$$

$$= 173.30 \text{ kW} \quad (170 \text{ kW})$$

The answer is (B).

Why Other Options Are Wrong

(A) This incorrect answer omits the efficiencies of the compressor and electric motor when calculating the power required. The power reported in this answer is the theoretical work required.

(C) This incorrect answer omits the number of stages in the exponent of the pressure ratio when calculating the power required.

(D) This incorrect answer uses an incorrect expression for the power required. The formula used was

$$P = 3F\left(\frac{k}{k-1}\right) RT_0 \left(\frac{p_3}{p_0}\right)^{\frac{k-1}{3k}} - 1$$

SOLUTION 30

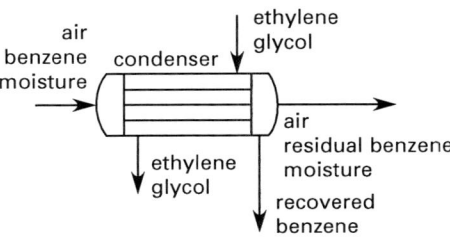

The removal efficiency, η, is given as 0.92. The benzene inlet volume fraction, y, is given as 0.35.

Calculate the air/benzene mixture mass flow rate in the inlet stream. The universal constant, R, is 10.73 lbf-ft^3/lbmol-in^2-°R. The temperature of the system is

$$T = 77°\text{F} + 460° = 537°\text{R}$$

The total pressure of the system, p, is 14.7 lbf/in^2. The specific volume of the air/benzene mixture is

$$v = \frac{RT}{p} = \frac{\left(10.73 \, \frac{\text{lbf-ft}^3}{\text{lbmol-in}^2\text{-°R}}\right)(537°\text{R})}{14.7 \, \frac{\text{lbf}}{\text{in}^2}}$$

$$= 392 \text{ ft}^3/\text{lbmol}$$

The inlet stream volumetric flow rate, $\dot{V}$, is given as 200 ft^3/min. The molar flow rate of benzene is

$$F = \left(\frac{\dot{V}}{v}\right) y = \left(\frac{200 \, \frac{\text{ft}^3}{\text{min}}}{392 \, \frac{\text{ft}^3}{\text{lbmol}}}\right)(0.35)\left(60 \, \frac{\text{min}}{\text{hr}}\right)$$

$$= 10.71 \text{ lbmol/hr}$$

The molar flow rate of benzene in the outlet stream is

$$F_{\text{out}} = F(1-\eta) = \left(10.71 \, \frac{\text{lbmol}}{\text{hr}}\right)(1-0.92)$$
$$= 0.857 \, \text{lbmol/hr}$$

The condensed benzene molar flow rate is

$$F_{\text{cond}} = F - F_{\text{out}} = 10.71 \, \frac{\text{lbmol}}{\text{hr}} - 0.857 \, \frac{\text{lbmol}}{\text{hr}}$$
$$= 9.853 \, \text{lbmol/hr}$$

The molecular weight of benzene, MW, is given as 78 lbm/lbmol. The mass flow rate of benzene recovered is

$$\dot{m} = F_{\text{cond}}(\text{MW}) = \left(9.853 \, \frac{\text{lbmol}}{\text{hr}}\right)\left(78 \, \frac{\text{lbm}}{\text{lbmol}}\right)$$
$$= 768.53 \, \text{lbm/hr} \quad (770 \, \text{lbm/hr})$$

The answer is (C).

Why Other Options Are Wrong

(A) This incorrect answer does not multiply by the molecular weight of benzene when calculating the molar flow rate of benzene in the outlet stream. Thus, the answer is reported in the wrong units.

(B) This incorrect answer does not use the conversion factor to convert from minutes to hours when calculating the molar flow rate of benzene.

(D) This incorrect answer uses a stream of pure benzene. This answer omits the inlet fraction of benzene of 0.35 when calculating the molar flow rate of benzene.

FLUIDS
SOLUTION 31

The length of straight pipe, L, is given as 10 ft. The elbow resistance coefficient, K_e, is given as 0.4. The valve resistance coefficient, K_v, is given as 5.6. The gravitational acceleration, g, is 32.17 ft/sec². The internal area of the pipe, A, is given as 0.0233 ft².

The volumetric flow rate is

$$Q = \left(100 \, \frac{\text{gal}}{\text{min}}\right)\left(\frac{1 \, \text{min}}{60 \, \text{sec}}\right)\left(\frac{1 \, \text{ft}^3}{7.48 \, \text{gal}}\right)$$
$$= 0.22282 \, \text{ft}^3/\text{sec}$$

The velocity of the fluid is

$$\text{v} = \frac{Q}{A} = \frac{0.22282 \, \frac{\text{ft}^3}{\text{sec}}}{0.0233 \, \text{ft}^2}$$
$$= 9.563 \, \text{ft/sec}$$

The head loss due to friction is

$$h_f = \left(\frac{17.4 \, \text{ft}}{100 \, \text{ft}}\right) L + \Sigma K \left(\frac{\text{v}^2}{2g}\right)$$
$$= \left(\frac{17.4 \, \text{ft}}{100 \, \text{ft}}\right) L + (K_e + K_v)\left(\frac{\text{v}^2}{2g}\right)$$
$$= \left(\frac{17.4 \, \text{ft}}{100 \, \text{ft}}\right)(10 \, \text{ft}) + (0.4 + 5.6)$$
$$\times \left(\frac{\left(9.563 \, \frac{\text{ft}}{\text{sec}}\right)^2}{(2)\left(32.17 \, \frac{\text{ft}}{\text{sec}^2}\right)}\right)$$
$$= 10.27 \, \text{ft} \quad (10 \, \text{ft})$$

The answer is (C).

Why Other Options Are Wrong

(A) This incorrect answer is the head loss due to friction of the pipe and elbow. This answer omits the friction loss through the valve.

(B) This incorrect answer is the head loss due to friction of the pipe and valve. This answer excludes the friction loss through the elbow.

(D) This incorrect answer omits the gravitational acceleration when calculating the head loss due to friction.

SOLUTION 32

The internal area of the pipe, A, is given as 0.0233 ft². The volumetric flow rate is

$$Q = \left(86 \, \frac{\text{gal}}{\text{min}}\right)\left(\frac{1 \, \text{min}}{60 \, \text{sec}}\right)\left(\frac{1 \, \text{ft}^3}{7.48 \, \text{gal}}\right)$$
$$= 0.19162 \, \text{ft}^3/\text{sec}$$

The velocity of the fluid is

$$\text{v} = \frac{Q}{A} = \frac{0.19162 \, \frac{\text{ft}^3}{\text{sec}}}{0.0233 \, \text{ft}^2}$$
$$= 8.224 \, \text{ft/sec}$$

The resistance coefficient of the elbow, K_e, is given as 0.4. The resistance coefficient of the valve, K_v, is given as 5.6. The atmospheric pressure, p, is given as 14.7 lbf/in². The gravitational acceleration, g, is 32.17 ft/sec². The gravitational constant, g_c, is 32.17 lbm-ft/lbf-sec². The density of the water at 170°F, ρ, is given as 62.4 lbm/ft³.

Let the top of the water layer be called point a′ and the pump suction be called point a. A mechanical energy balance from point a′ to point a with no pump work is

$$\frac{p_{a'}}{\rho} + \left(\frac{g}{g_c}\right)z_{a'} + \frac{v_{a'}^2}{2g_c}$$
$$= \frac{p_a}{\rho} + \left(\frac{g}{g_c}\right)z_a + \frac{v_a^2}{2g_c} + \left(\frac{g}{g_c}\right)h_f$$

Solving for p_a/ρ,

$$\frac{p_a}{\rho} = \frac{p_{a'}}{\rho} + \left(\frac{g}{g_c}\right)(z_{a'} - z_a) + \frac{v_{a'}^2}{2g_c} - \frac{v_a^2}{2g_c} - \left(\frac{g}{g_c}\right)h_f$$

Let the net positive suction head available be the pressure head available at point a above the vapor pressure head.

$$\text{NPSHA} \equiv \left(\frac{g_c}{g}\right)\left(\frac{p_a - p_v}{\rho}\right)$$

Eliminating p_a and simplifying,

$$\text{NPSHA} = \frac{g_c p_{a'}}{g\rho} - \frac{g_c p_v}{g\rho} - h_f + z_{a'} - z_a + \frac{v_{a'}^2}{2g} - \frac{v_a^2}{2g}$$

The pressure at point a′ is given as

$$(p_{a'})_{\text{vac}} = \frac{\Delta z_{a'}\rho g}{g_c}$$

$$= \frac{(20 \text{ in})\left(\frac{1 \text{ ft}}{12 \text{ in}}\right)(13.61)\left(62.4 \frac{\text{lbm}}{\text{ft}^3}\right)}{32.17 \frac{\text{lbm-ft}}{\text{lbf-sec}^2}}$$

$$= 9.829 \text{ lbf/in}^2$$

The absolute pressure at point a′ is

$$p_{a'} = p - (p_{a'})_{\text{vac}} = 14.7 \frac{\text{lbf}}{\text{in}^2} - 9.829 \frac{\text{lbf}}{\text{in}^2}$$
$$= 4.871 \text{ lbf/in}^2$$

The length of the suction line, L, is given as 12 ft. The friction head loss is

$$\left(\frac{g_c}{g}\right)h_f = \left(\frac{16.4 \text{ ft}}{100 \text{ ft}}\right)L + \Sigma K\left(\frac{v^2}{2g}\right)$$

$$= \left(\frac{16.4 \text{ ft}}{100 \text{ ft}}\right)(12 \text{ ft}) + (0.4 + 5.6)$$
$$\times \left(\frac{\left(8.224 \frac{\text{ft}}{\text{sec}}\right)^2}{(2)\left(32.17 \frac{\text{ft}}{\text{sec}^2}\right)}\right)$$

$$= 8.28 \text{ ft}$$

Let the datum at point a, Z_a, be defined as 0 ft. Assume that the velocity heads are negligible.

$$v_{a'} \approx v_a \approx 0 \text{ ft/sec}$$

$$\text{NPSHA} = \frac{g_c p_{a'}}{g\rho} + z_{a'} - z_a - \frac{g_c p_v}{g\rho} - h_f$$

$$= \frac{\left(32.17 \frac{\text{lbm-ft}}{\text{lbf-sec}^2}\right)\left(4.871 \frac{\text{lbf}}{\text{in}^2}\right)\left(144 \frac{\text{in}^2}{\text{ft}^2}\right)}{\left(32.17 \frac{\text{ft}}{\text{sec}^2}\right)\left(62.4 \frac{\text{lbm}}{\text{ft}^3}\right)}$$

$$+ 16 \text{ ft} - 0 \text{ ft}$$

$$- \frac{\left(32.17 \frac{\text{lbm-ft}}{\text{lbf-sec}^2}\right)\left(6.0 \frac{\text{lbf}}{\text{in}^2}\right)\left(144 \frac{\text{in}^2}{\text{ft}^2}\right)}{\left(32.17 \frac{\text{ft}}{\text{sec}^2}\right)\left(62.4 \frac{\text{lbm}}{\text{ft}^3}\right)}$$

$$- 8.27 \text{ ft}$$
$$= 5.115 \text{ ft} \quad (5.0 \text{ ft})$$

The answer is (A).

Why Other Options Are Wrong

(B) This incorrect answer adds the head loss due to friction instead of subtracting it when calculating the net positive suction head available.

(C) This incorrect answer adds the pressure head due to the water vapor instead of subtracting it when calculating the net positive suction head available.

(D) This incorrect answer uses the opposite sign for the head due to gauge pressure when calculating the pressure at a′.

SOLUTION 33

The water density, ρ_w, is given as 62.4 lbm/ft³. The hydrocarbon layer density, ρ_{HC}, is given as 43.68 lbm/ft³. The average liquid velocity at the pump suction nozzle, v, is given as approximately 0 ft/sec. The head loss due to friction in the suction line, h_f, is given as 2.8 ft. The bottom of the separator height, h_B, is given as 1 ft. The vapor pressure of the liquid at the pump inlet, p_v, is given as 1.70 lbf/in². The gravitational acceleration, g, is 32.17 ft/sec². The gravitational constant, g_c, is 32.17 lbm-ft/lbf-sec².

Let the top of the hydrocarbon layer be called point a′, the top of water layer be called point b, and the pump suction be called point a. A mechanical energy balance from point a′ to point b with no pump work and no friction is

$$\frac{p_{a'}}{\rho_{\text{HC}}} + \left(\frac{g}{g_c}\right)z_{a'} + \frac{v_{a'}^2}{2g_c}$$
$$= \frac{p_b}{\rho_{\text{HC}}} + \left(\frac{g}{g_c}\right)z_b + \frac{v_b^2}{2g_c}$$

A mechanical energy balance from point b to point a with no pump work is

$$\frac{p_b}{\rho_w} + \left(\frac{g}{g_c}\right) z_b + \frac{v_b^2}{2g_c}$$
$$= \frac{p_a}{\rho_w} + \left(\frac{g}{g_c}\right) z_a + \frac{v_a^2}{2g_c} + \left(\frac{g}{g_c}\right) h_f$$

Solving for p_b/ρ_{HC},

$$\frac{p_b}{\rho_{HC}} = \frac{p_{a'}}{\rho_{HC}} + \left(\frac{g}{g_c}\right)(z_b - z_a) + \frac{v_b^2 - v_a^2}{2g_c}$$
$$- \left(\frac{g}{g_c}\right) h_f$$

Eliminating p_b,

$$\frac{p_a}{\rho_w} = \left(\frac{\rho_{HC}}{\rho_w}\right)\left(\frac{p_{a'}}{\rho_{HC}} + \left(\frac{g}{g_c}\right)(z_{a'} - z_b) + \frac{v_{a'}^2 - v_b^2}{2g_c}\right)$$
$$+ \left(\frac{g}{g_c}\right)(z_b - z_a) + \left(\frac{v_b^2 - v_a^2}{2g_c}\right) - \left(\frac{g}{g_c}\right) h_f$$

Let the net positive suction head available be the pressure head at point a above the vapor pressure head.

$$\text{NPSHA} = \left(\frac{g_c}{g}\right)\left(\frac{p_a - p_v}{\rho_w}\right)$$

Eliminating p_a gives

$$\text{NPSHA} = \left(\frac{g}{g_c}\right)\left(\frac{p_{a'} - p_v}{\rho_w}\right) + \left(\frac{\rho_{HC}}{\rho_w}\right)(z_{a'} - z_b)$$
$$+ z_b - z_a + \left(\frac{\rho_{HC}}{\rho_w}\right)\left(\frac{v_{a'}^2 - v_b^2}{2g}\right)$$
$$+ \frac{v_b^2 - v_a^2}{2g} - h_f$$

Let the datum at point a, z_a, be defined as 0 ft, and let $v_{a'}^2 = v_b^2 \approx 0 \text{ ft}^2/\text{sec}^2$ and $v_a^2 \approx 0 \text{ ft}^2/\text{sec}^2$.

$$\text{NPSHA} = \left(\frac{g_c}{g}\right)\left(\frac{p_{a'} - p_v}{\rho_w}\right)$$
$$+ \left(\frac{\rho_{HC}}{\rho_w}\right) z_{a'} - z_b + z_b - z_a$$
$$+ \left(\frac{\rho_{HC}}{\rho_w}\right)\left(\frac{v_{a'}^2 - v_b^2}{2g}\right) + \frac{v_b^2 - v_a^2}{2g} - h_f$$

$$= \left(\frac{32.17 \frac{\text{lbm-ft}}{\text{lbf-sec}^2}}{32.17 \frac{\text{ft}}{\text{sec}^2}}\right) \left(\frac{\left(21.6 \frac{\text{lbf}}{\text{in}^2} - 1.7 \frac{\text{lbf}}{\text{in}^2}\right) \times \left(144 \frac{\text{in}^2}{\text{ft}^2}\right)}{62.4 \frac{\text{lbm}}{\text{ft}^3}}\right)$$

$$+ \left(\frac{43.68 \frac{\text{lbm}}{\text{ft}^3}}{62.4 \frac{\text{lbm}}{\text{ft}^3}}\right)(5.5 \text{ ft} - 2 \text{ ft})$$

$$+ 2 \text{ ft} - 0 \text{ ft} + \left(\frac{43.68 \frac{\text{lbm}}{\text{ft}^3}}{62.4 \frac{\text{lbm}}{\text{ft}^3}}\right)$$
$$\times \left(\frac{\left(0 \frac{\text{ft}}{\text{sec}}\right)^2 - \left(0 \frac{\text{ft}}{\text{sec}}\right)^2}{(2)\left(32.17 \frac{\text{ft}}{\text{sec}^2}\right)}\right)$$

$$+ \frac{\left(0 \frac{\text{ft}}{\text{sec}}\right)^2 - \left(0 \frac{\text{ft}}{\text{sec}}\right)^2}{(2)\left(32.17 \frac{\text{ft}}{\text{sec}^2}\right)} - 2.8 \text{ ft}$$

$$= 47.573 \text{ ft} \quad (48 \text{ ft})$$

The answer is (C).

Why Other Options Are Wrong

(A) This incorrect answer omits the friction head when calculating the net positive head available.

(B) This incorrect answer excludes the static head when calculating the net positive suction head available.

(D) This incorrect answer uses the opposite sign for the friction losses head when calculating the net positive head available.

SOLUTION 34

Let the top of the hydrocarbon layer be called point a', the top of the water layer be called point b, and the pump suction be called point a. A mechanical energy balance from point a' to point b with no pump work and no friction is

$$\frac{p_{a'}}{\rho_{HC}} + \left(\frac{g}{g_c}\right) z_{a'} + \frac{v_{a'}^2}{2g_c} = \frac{p_b}{\rho_{HC}} + \left(\frac{g}{g_c}\right) z_b + \frac{v_b^2}{2g_c}$$

A mechanical energy balance from point b to point a with no pump work is

$$\frac{p_b}{\rho_w} + \left(\frac{g}{g_c}\right) z_b + \frac{v_b^2}{2g_c}$$
$$= \frac{p_a}{\rho_w} + \left(\frac{g}{g_c}\right) z_a + \frac{v_a^2}{2g_c} + \left(\frac{g}{g_c}\right) h_f$$

Solving for p_b/ρ_{HC},

$$\frac{p_b}{\rho_{HC}} = \frac{p_{a'}}{\rho_{HC}} + \left(\frac{g}{g_c}\right)(z_b - z_a) + \frac{v_b^2 - v_a^2}{2g_c} - \left(\frac{g}{g_c}\right) h_f$$

Eliminating p_b,

$$\frac{p_a}{\rho_w} = \left(\frac{\rho_{HC}}{\rho_w}\right)\left(\frac{p_{a'}}{\rho_{HC}} + \left(\frac{g}{g_c}\right)(z_{a'} - z_b) + \frac{v_{a'}^2 - v_b^2}{2g_c}\right)$$
$$+ \left(\frac{g}{g_c}\right)(z_b - z_a) + \left(\frac{v_b^2 - v_a^2}{2g_c}\right) - \left(\frac{g}{g_c}\right) h_f$$

The pump suction head is made of static, dynamic, and friction heads.

$$h_s = \frac{g p_{a'}}{g_c \rho_w} + \left(\frac{\rho_{HC}}{g \rho_w}\right)(z_{a'} - z_b) + z_b - z_a$$

So a decrease in ρ_{HC} will decrease the pump static suction head.

The answer is (C).

Why Other Options Are Wrong

(A) This answer is incorrect because, from the equation for pump suction head, a decrease in ρ_{HC} will decrease the pump suction head.

(B) This answer is incorrect because if ρ_{HC} were to decrease, the velocity head would decrease.

(D) This answer is incorrect because if ρ_{HC} were to decrease, the pump suction head, the flow through the pipe, and the frictional head loss would all decrease.

SOLUTION 35

The viscosity of the fluid is

$$\mu = (0.9 \text{ cP}) \left(\frac{0.000672 \frac{\text{lbm}}{\text{ft-sec}}}{1 \text{ cP}}\right)$$
$$= 0.000605 \text{ lbm/ft-sec}$$

The gravitational acceleration, g, is 32.17 ft/sec^2. The density of the fluid, ρ, is given as 61.8 lbm/ft^3. Because the pipe is a 4 in schedule-40 steel pipe, the pipe diameter is

$$D = (4.026 \text{ in}) \left(\frac{1 \text{ ft}}{12 \text{ in}}\right) = 0.3355 \text{ ft}$$

The volumetric flow rate is

$$Q = \left(330 \frac{\text{gal}}{\text{min}}\right)\left(\frac{1 \text{ min}}{60 \text{ sec}}\right)\left(\frac{1 \text{ ft}^3}{7.48 \text{ gal}}\right)$$
$$= 0.7353 \text{ ft}^3/\text{sec}$$

The pipe inner area is

$$A = \tfrac{\pi}{4} D^2 = \left(\frac{\pi}{4}\right)(0.3355 \text{ ft})^2$$
$$= 0.0884 \text{ ft}^2$$

The flow velocity is

$$v = \frac{Q}{A} = \frac{0.7353 \frac{\text{ft}^3}{\text{sec}}}{0.0884 \text{ ft}^2}$$
$$= 8.32 \text{ ft/sec}$$

The Reynolds number is

$$\text{Re} = \frac{D \rho v}{\mu}$$
$$= \frac{(0.3355 \text{ ft})\left(61.8 \frac{\text{lbm}}{\text{ft}^3}\right)\left(8.32 \frac{\text{ft}}{\text{sec}}\right)}{0.000605 \frac{\text{lbm}}{\text{ft-sec}}}$$
$$= 285{,}134$$

The pipe roughness, ϵ, is given as 0.000144 ft. The relative pipe roughness is

$$\frac{\epsilon}{D} = \frac{0.000144 \text{ ft}}{0.3355 \text{ ft}} = 0.00043$$

The friction factor, f, from the Moody diagram is 0.018. The equivalent length of pump discharge line, L_e, is given as 483 ft. The head loss due to friction is

$$h_f = \frac{f L_e v^2}{2 D g} = \frac{(0.018)(483 \text{ ft})\left(8.32 \frac{\text{ft}}{\text{sec}}\right)^2}{(2)(0.3355 \text{ ft})\left(32.17 \frac{\text{ft}}{\text{sec}^2}\right)}$$
$$= 27.88 \text{ ft} \quad (28 \text{ ft})$$

The answer is (C).

Why Other Options Are Wrong

(A) This incorrect answer uses the pipe diameter in inches instead of feet when calculating the line pressure drop.

(B) This incorrect answer omits the exponent in the velocity term when calculating the line pressure drop.

(D) This incorrect answer erroneously omits the factor of 2 in the denominator of the expression for the line pressure drop.

SOLUTION 36

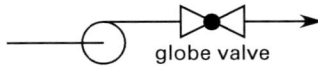
globe valve

The pipe internal diameter is given as

$$D = (2.067 \text{ in})\left(\frac{1 \text{ ft}}{12 \text{ in}}\right) = 0.17225 \text{ ft}$$

The design flow rate is given as

$$Q = \left(345 \frac{\text{gal}}{\text{min}}\right)\left(\frac{1 \text{ min}}{60 \text{ sec}}\right)\left(\frac{1 \text{ ft}^3}{7.48 \text{ gal}}\right)$$
$$= 0.7687 \text{ ft}^3/\text{sec}$$

The pipe internal area is

$$A = \tfrac{\pi}{4}D^2 = \left(\tfrac{\pi}{4}\right)(0.17225 \text{ ft})^2$$
$$= 0.0233 \text{ ft}^2$$

The flow velocity is

$$v = \frac{Q}{A} = \frac{0.7687 \dfrac{\text{ft}^3}{\text{sec}}}{0.0233 \text{ ft}^2}$$
$$= 32.99 \text{ ft/sec}$$

The equivalent length of open globe valve, $L_{e,B}$, is given as 110 ft. The gravitational acceleration, g, is 32.17 ft/sec^2. The gravitational constant, g_c, is 32.17 lbm-ft/lbf-sec^2. The friction factor, f, is given as 0.029213. The head loss due to friction across the open control valve is

$$h_f = \frac{fL_{e,B}v^2}{2Dg}$$

$$= \frac{(0.029213)(110 \text{ ft})\left(32.99 \dfrac{\text{ft}}{\text{sec}}\right)^2}{(2)(0.17225 \text{ ft})\left(32.17 \dfrac{\text{ft}}{\text{sec}^2}\right)}$$

$$= 315.568 \text{ ft}$$

The density of the fluid, ρ_L, is given as 61.8 lbm/ft^3. The density of pure water at 60°F, ρ, is 62.4 lbm/ft^3. The pressure drop head across the open control valve is

$$\Delta p = \frac{\rho g h_f}{g_c}$$

$$= \frac{\left(61.8 \dfrac{\text{lbm}}{\text{ft}^3}\right)\left(32.17 \dfrac{\text{ft}}{\text{sec}^2}\right) \times (315.57 \text{ ft})\left(\dfrac{1 \text{ ft}^2}{144 \text{ in}^2}\right)}{32.17 \dfrac{\text{lbm-ft}}{\text{lbf-sec}^2}}$$

$$= 135.432 \text{ lbf/in}^2$$

The specific gravity of the fluid is the ratio of the density of the fluid to the density of water at 60°F.

$$SG = \frac{\rho_L}{\rho} = \frac{61.8 \dfrac{\text{lbm}}{\text{ft}^3}}{62.4 \dfrac{\text{lbm}}{\text{ft}^3}} = 0.9904$$

$$C_v = \frac{Q \dfrac{\text{gal}}{\text{min}}}{\sqrt{\dfrac{\Delta p \dfrac{\text{lbf}}{\text{in}^2}}{SG}}} = \frac{345 \dfrac{\text{gal}}{\text{min}}}{\sqrt{\dfrac{135.432 \dfrac{\text{lbf}}{\text{in}^2}}{0.9904}}}$$

$$= \frac{345 \dfrac{\text{gal}}{\text{min}}}{\sqrt{136.745 \dfrac{\text{lbf}}{\text{in}^2}}} \quad \left(345 \dfrac{\dfrac{\text{gal}}{\text{min}}}{137 \dfrac{\text{lbf}}{\text{in}^2}}\right)$$

The answer is (C).

Why Other Options Are Wrong

(A) This incorrect answer uses both the pipe discharge line equivalent length and the equivalent length of the globe valve when calculating the head loss due to friction.

(B) When calculating the pressure head, this incorrect answer uses the equivalent length of the pipe instead of the equivalent length of the globe valve.

(D) This incorrect answer omits the specific gravity of the fluid when calculating the flow coefficient.

SOLUTION 37

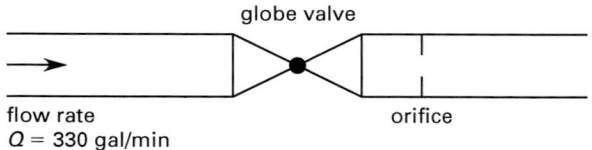
flow rate
Q = 330 gal/min
globe valve
orifice

The pipe diameter is given as

$$D = (4.026 \text{ in})\left(\frac{1 \text{ ft}}{12 \text{ in}}\right) = 0.3355 \text{ ft}$$

The pipe internal area is

$$A = \tfrac{\pi}{4}D^2 = \left(\frac{\pi}{4}\right)(0.3355 \text{ ft})^2$$
$$= 0.0884 \text{ ft}^2$$

The flow velocity is

$$v = \frac{Q}{A} = \frac{\left(330 \frac{\text{gal}}{\text{min}}\right)\left(\frac{1 \text{ min}}{60 \text{ sec}}\right)\left(\frac{1 \text{ ft}^3}{7.48 \text{ gal}}\right)}{0.0884 \text{ ft}^2}$$
$$= 8.32 \text{ ft/sec}$$

The ratio of the orifice diameter to the inlet diameter is

$$\beta = \frac{D_o}{D} = 0.68$$

The orifice coefficient, C_o, is given as 0.60. The gravitational acceleration, g, is 32.17 ft/sec². The flow through the orifice plate is

$$Q_o = \frac{C_o A_o}{\sqrt{1-\beta^4}}\sqrt{2gh_o}$$

The volumetric flow rate is

$$Q_o = Q = Av$$

Combining,

$$Av = \frac{C_o A_o}{\sqrt{1-\beta^4}}\sqrt{2gh_o}$$

As A_o/A equals β^2, we can solve for the head loss of the orifice plate.

$$h_o = \left(\frac{v^2}{2gC_o^2}\right)\left(\frac{1-\beta^4}{\beta^4}\right)$$
$$= \left(\frac{\left(8.32 \frac{\text{ft}}{\text{sec}}\right)^2}{(2)\left(32.17 \frac{\text{ft}}{\text{sec}^2}\right)(0.60)^2}\right)\left(\frac{1-(0.68)^4}{(0.68)^4}\right)$$
$$= 10.9819 \text{ ft}$$

The permanent pressure drop is given as 54%. The pressure drop through the orifice plate is

$$\Delta p_o = h_o \Delta p = (10.9819 \text{ ft})(0.54)$$
$$= 5.9302 \text{ ft} \quad (5.9 \text{ ft})$$

The answer is (B).

Why Other Options Are Wrong

(A) This incorrect answer leaves out the orifice coefficient when calculating the pressure drop of the orifice plate.

(C) This incorrect answer excludes the permanent pressure drop when calculating the pressure drop through the orifice plate.

(D) This answer erroneously divides the scale reading of the orifice plate by the permanent pressure drop instead of multiplying the two terms when calculating the pressure drop through the orifice plate.

SOLUTION 38

The diameter of the pipe is given as

$$D = (6.065 \text{ in})\left(\frac{1 \text{ ft}}{12 \text{ in}}\right) = 0.5054 \text{ ft}$$

The viscosity of the fluid is given as

$$\mu = (470 \text{ cP})\left(\frac{6.72 \times 10^{-4} \frac{\text{lbm}}{\text{ft-sec}}}{1 \text{ cP}}\right)$$
$$= 0.3158 \text{ lbm/ft-sec}$$

The cross-sectional area of the pipe is

$$A = \tfrac{\pi}{4}D^2 = \left(\frac{\pi}{4}\right)(0.5054 \text{ ft})^2$$
$$= 0.2006 \text{ ft}^2$$

The liquid volumetric flow rate is

$$Q = \left(800 \frac{\text{gal}}{\text{min}}\right)\left(\frac{1 \text{ ft}^3}{7.48 \text{ gal}}\right)\left(\frac{1 \text{ min}}{60 \text{ sec}}\right)$$
$$= 1.7825 \text{ ft}^3/\text{sec}$$

The velocity of the liquid is

$$v_2 = \frac{Q}{A} = \frac{1.7825 \frac{\text{ft}^3}{\text{sec}}}{0.2006 \text{ ft}^2}$$
$$= 8.886 \text{ ft/sec}$$

The density, ρ, of the liquid is given as 56.13 lbm/ft³. The Reynolds number is

$$\text{Re} = \frac{Dv\rho}{\mu}$$
$$= \frac{(0.5054 \text{ ft})\left(8.886 \frac{\text{ft}}{\text{sec}}\right)\left(56.13 \frac{\text{lbm}}{\text{ft}^3}\right)}{0.3158 \frac{\text{lbm}}{\text{ft-sec}}}$$
$$= 798.22$$

Because the Reynolds number is less than 2100, the flow is laminar. For laminar flow, the Darcy friction factor is

$$f = \frac{64}{\text{Re}} = \frac{64}{798.22} = 0.08018$$

The length of the pipe is

$$L = 225 \text{ ft} + 100 \text{ ft} + 20 \text{ ft}$$
$$= 345 \text{ ft}$$

The acceleration due to gravity, g, is 32.17 ft/sec². The gravitational constant, g_c, is 32.17 lbm-ft/lbf-sec². The pressure at point 2, p_2, is given as 15 lbf/in². The general mechanical energy balance for flow from point 1 to point 2 with pump work and friction is

$$\frac{p_1}{\rho} + \left(\frac{g}{g_c}\right) z_1 + \frac{v_1^2}{2g_c} + \eta W_p$$
$$= \frac{p_2}{\rho} + \left(\frac{g}{g_c}\right) z_2 + \frac{v_2^2}{2g_c} + \left(\frac{g}{g_c}\right) h_f$$

The mechanical energy balance from point 1 to point 2 with no pump work and no velocity change is

$$p_1 - p_2 + \left(\frac{g\rho}{g_c}\right)(z_1 - z_2) = \left(\frac{g\rho}{g_c}\right) h_f \quad [\text{I}]$$

Darcy's law is

$$h_f = \left(\frac{fL}{D} + \sum K\right)\left(\frac{v_2^2}{2g}\right)$$
$$= \left(\frac{(0.08018)(345 \text{ ft})}{0.5054 \text{ ft}} + 0.30 + 0.12 + 2.25\right)$$
$$\times \left(\frac{\left(8.886 \dfrac{\text{ft}}{\text{sec}}\right)^2}{(2)\left(32.17 \dfrac{\text{ft}}{\text{sec}^2}\right)}\right)$$
$$= 70.448 \text{ ft}$$

Replacing z_1 with 0 ft in Eq. I gives

$$p_1 - 15 \frac{\text{lbf}}{\text{in}^2} + \left(\frac{\left(32.17 \dfrac{\text{ft}}{\text{sec}^2}\right)\left(56.13 \dfrac{\text{lbm}}{\text{ft}^3}\right)}{32.17 \dfrac{\text{lbm-ft}}{\text{lbf-sec}^2}}\right)$$
$$\times \left(\frac{1 \text{ ft}^2}{144 \text{ in}^2}\right)(0 \text{ ft} - 20 \text{ ft})$$
$$= \left(\frac{\left(32.17 \dfrac{\text{ft}}{\text{sec}^2}\right)\left(56.13 \dfrac{\text{lbm}}{\text{ft}^3}\right)}{32.17 \dfrac{\text{lbm-ft}}{\text{lbf-sec}^2}}\right)$$
$$\times \left(\frac{1 \text{ ft}^2}{144 \text{ in}^2}\right)(70.448 \text{ ft})$$

$$p_1 - 15 \frac{\text{lbf}}{\text{in}^2} - 7.7958 \frac{\text{lbf}}{\text{in}^2} = 27.46 \frac{\text{lbf}}{\text{in}^2}$$

Solving for p_1 gives

$$p_1 = 15 \frac{\text{lbf}}{\text{in}^2} + 7.7958 \frac{\text{lbf}}{\text{in}^2} + 27.46 \frac{\text{lbf}}{\text{in}^2}$$
$$= 50.256 \text{ lbf/in}^2 \quad (50 \text{ lbf/in}^2)$$

The answer is (C).

Why Other Options Are Wrong

(A) This incorrect answer calculates the pressure at point 1 by subtracting the pressure at point 2 from the pressure difference instead of adding the pressure at point 2 to the pressure difference.

(B) This incorrect answer uses the diameter of the pipe in inches instead of feet when calculating the head loss due to friction.

(D) This incorrect answer does not use the conversion factor from square inches to square feet when calculating the pressure difference due to elevation.

SOLUTION 39

The diameter of the pipe is given as

$$D = (1.049 \text{ in})\left(\frac{1 \text{ ft}}{12 \text{ in}}\right) = 0.08742 \text{ ft}$$

The pressure drop due to friction is given as

$$\Delta p_f = \left(26.35 \frac{\text{lbf}}{\text{in}^2}\right)\left(144 \frac{\text{in}^2}{\text{ft}^2}\right)$$
$$= 3794.4 \text{ lbf/ft}^2$$

The velocity, v, is given as 10 ft/sec. The density of the fluid, ρ, is given as 123.6 lbm/ft³. The length of the pipe, L, is given as 54 ft. The roughness of the pipe, ϵ, is given as 0.00015 ft. The gravitational constant, g_c, is 32.17 lbm-ft/lbf-sec². The friction factor, from the Darcy equation, is

$$f = \frac{2\Delta p_f D g_c}{\rho v^2 L}$$
$$= \frac{(2)\left(3794.4 \dfrac{\text{lbf}}{\text{ft}^2}\right)(0.08742 \text{ ft})\left(32.17 \dfrac{\text{lbm-ft}}{\text{lbf-sec}^2}\right)}{\left(123.6 \dfrac{\text{lbm}}{\text{ft}^3}\right)\left(10 \dfrac{\text{ft}}{\text{sec}}\right)^2 (54 \text{ ft})}$$
$$= 0.031975$$

Calculate the Reynolds number using the Colebrook and White equation.

$$\frac{1}{\sqrt{f}} = -2\log\left(\frac{\epsilon}{3.7D} + \frac{2.51}{\text{Re}\sqrt{f}}\right)$$

Solving the preceding equation for the Reynolds number gives

$$\text{Re} = \cfrac{2.51}{\left(10^{\frac{-1}{2\sqrt{f}}} - \cfrac{\epsilon}{3.7D}\right)\sqrt{f}}$$

$$= \cfrac{2.51}{\left(10^{\frac{-1}{2\sqrt{0.031975}}} - \cfrac{0.00015 \text{ ft}}{(3.7)(0.08742 \text{ ft})}\right)}$$
$$\times \sqrt{0.031975}$$

$$= 12{,}365$$

The flow is turbulent because the Reynolds number is larger than 2100.

From the definition of the Reynolds number, the viscosity of the fluid is

$$\mu = \frac{\rho D \text{v}}{\text{Re}}$$

$$= \left(\frac{\left(123.6 \ \frac{\text{lbm}}{\text{ft}^3}\right)(0.08742 \text{ ft})\left(10 \ \frac{\text{ft}}{\text{sec}}\right)}{12{,}365}\right)$$

$$\times \left(\frac{1 \text{ cP}}{0.000672 \ \frac{\text{lbm}}{\text{ft-sec}}}\right)$$

$$= 13.0037 \text{ cP} \quad (13 \text{ cP})$$

The answer is (B).

Why Other Options Are Wrong

(A) This incorrect answer omits the conversion factor from lbm/ft-sec to centipoise when calculating the viscosity.

(C) This incorrect answer calculates the Reynolds number using the expression for smooth pipes.

(D) This incorrect answer uses the Poiseuille equation to calculate the viscosity. This equation applies to laminar flow, not turbulent flow.

SOLUTION 40

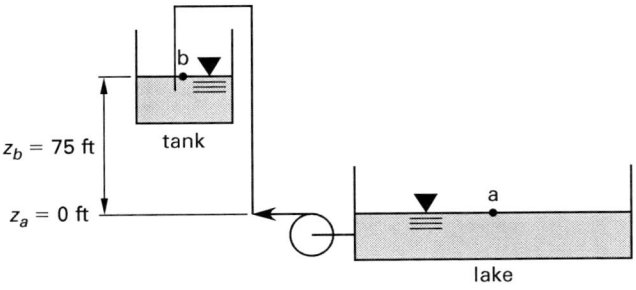

The pressure head, Δh_p, is given as 0 ft. The static head, Δh_s, is given as 75 ft. The pump head, h_{pump}, is given as 107 ft. The equivalent length of pipe, L, is given as 200 ft. The diameter of the pipe is given as

$$D = (2.067 \text{ in})\left(\frac{1 \text{ ft}}{12 \text{ in}}\right) = 0.17225 \text{ ft}$$

The gravitational acceleration, g, is 32.17 ft/sec^2. The gravitational constant, g_c, is 32.17 lbm-ft/lbf-sec^2.

Let the datum at point a, z_a, be defined as 0 ft. Points a and b are at atmospheric pressure, so $p_a = p_b$. The general mechanical energy balance for flow from point a to point b with pump work and friction is

$$\frac{p_a}{\rho} + \left(\frac{g}{g_c}\right)z_a + \frac{\text{v}_a^2}{2g_c} + \eta W_p$$
$$= \frac{p_b}{\rho} + \left(\frac{g}{g_c}\right)z_b + \frac{\text{v}_b^2}{2g_c} + \left(\frac{g}{g_c}\right)h_f$$

The mechanical energy balance from point a to point b with no velocity change is

$$h_f = \left(\frac{g_c}{g}\right)(p_a - p_b) + z_a - z_b + \left(\frac{g_c}{g}\right)\eta W_p$$
$$= 0 \text{ ft} + 0 \text{ ft} - 75 \text{ ft} + 107 \text{ ft}$$
$$= 32 \text{ ft} \qquad [\text{I}]$$

The friction loss term according to Darcy's law is

$$\left(\frac{g}{g_c}\right)h_f = \frac{fL\text{v}^2}{2Dg_c}$$

The continuity equation gives

$$\text{v} = \frac{Q}{A} = \frac{4Q}{\pi D^2}$$

Eliminating velocity gives

$$h_f = \frac{8LfQ^2}{\pi^2 g D^5}$$

$$= \left(\cfrac{(8)(200 \text{ ft})fQ^2}{\pi^2 \left(32.17 \ \frac{\text{ft}}{\text{sec}^2}\right)(0.17225 \text{ ft})^5}\right)$$

$$\times \left(\frac{1 \text{ min}}{60 \text{ sec}}\right)^2 \left(\frac{1 \text{ ft}^3}{7.48 \text{ gal}}\right)$$

$$= \left(0.16499 \ \cfrac{\text{ft}}{\left(\frac{\text{gal}}{\text{min}}\right)^2}\right)fQ^2 \qquad [\text{II}]$$

Equation I equals Eq. II.

$$32 \text{ ft} = \left(0.16499 \frac{\text{ft}}{\left(\frac{\text{gal}}{\text{min}}\right)^2}\right) fQ^2$$

$$f = \frac{32 \text{ ft}}{\left(0.16499 \frac{\text{ft}}{\left(\frac{\text{gal}}{\text{min}}\right)^2}\right) Q^2} \quad \text{[III]}$$

The density of water, ρ, is given as 62.43 lbm/ft^3. The viscosity of water, μ, is given as 1.42 cP. The Reynolds number is

$$\text{Re} = \frac{\rho Q \text{v}}{\mu} = \frac{\rho D 4 Q}{\mu \pi D^2} = \frac{4 \rho Q}{\pi D \mu}$$

$$= \frac{(4)\left(62.43 \frac{\text{lbm}}{\text{ft}^3}\right) Q \left(\frac{1 \text{ min}}{60 \text{ sec}}\right)\left(\frac{1 \text{ ft}^3}{7.48 \text{ gal}}\right)}{\pi(2.067 \text{ in})\left(\frac{1 \text{ ft}}{12 \text{ in}}\right)(1.42 \text{ cP})\left(\frac{0.000672 \frac{\text{lbm}}{\text{ft-sec}}}{\text{cP}}\right)}$$

$$= \left(1078 \frac{\text{min}}{\text{gal}}\right) Q$$

The friction factor using the Sacham equation is

$$f = \left(-2 \log \left(\frac{\frac{\epsilon}{D}}{3.7} - \frac{5.02}{\text{Re}} \log\left(\frac{\frac{\epsilon}{D}}{3.7} + \frac{14.5}{\text{Re}}\right)\right)\right)^{-2}$$

The roughness of the pipe, ϵ, is given as 0.00015 ft.

$$\frac{\text{pipe roughness}}{\text{pipe internal diameter}} = \frac{\epsilon}{D} = \frac{0.00015 \text{ ft}}{0.17225 \text{ ft}}$$

$$= 0.00087$$

Replacing,

$$f = \left(-2 \log \left(\frac{0.00087}{3.7} - \frac{5.02}{1078Q} \times \log\left(\frac{0.00087}{3.7} + \frac{14.5}{1078Q}\right)\right)\right)^{-2}$$

Simplifying,

$$f = \left(-2 \log \left(2.35 \times 10^{-4} - \frac{0.0047}{Q} \times \log\left(2.35 \times 10^{-4} + \frac{0.1345}{Q}\right)\right)\right)^{-2} \quad \text{[IV]}$$

Solving Eqs. III and IV simultaneously using a programmable calculator gives

$$Q = 94.495 \text{ gal/min} \quad (90 \text{ gal/min})$$

The answer is (A).

Alternate Solution

Equations III and IV could also be resolved iteratively by testing values of Q. Find the value of f from Eqs. III and IV. The value of Q that makes the value of f from Eq. III equal to the value of f from Eq. IV is the solution.

Q	Eq. III	Eq. IV
30	0.2155	0.0241
60	0.0539	0.0222
90	0.0239	0.0214
95.5	0.0213	0.0213

Why Other Options Are Wrong

(B) This answer erroneously calculates the friction factor as $h_f = 200 \text{ ft} - 107 \text{ ft} = 93 \text{ ft}$.

(C) This incorrect answer omits the static head when calculating the friction loss.

(D) This incorrect answer uses a negative static head when calculating the friction head.

SOLUTION 41

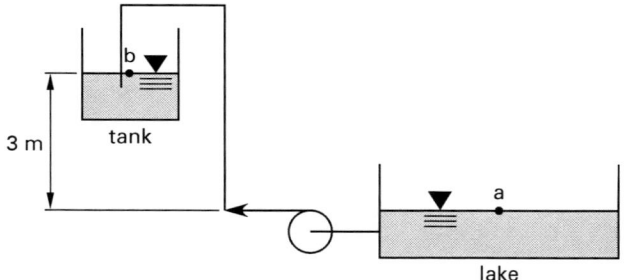

The density of water at 20°C, ρ, is given as 998.2 kg/m^3. The diameter of the pipe, D, is given as 0.0525 m. The internal area of the pipe, A, is given as 0.002165 m^2. The equivalent length of pipe, L, is given as 50 m. The gravitational acceleration, g, is 9.8 m/s^2. The gravitational constant, g_c, is 1 kg·m/N·s^2. The viscosity of the fluid, μ, is given as 1 cP. The roughness of the pipe is given as

$$\epsilon = (0.0457 \text{ mm})\left(\frac{1 \text{ m}}{1000 \text{ mm}}\right)$$

$$= 0.0000457 \text{ m}$$

The ratio of the pipe roughness to the pipe diameter is
$$\frac{\epsilon}{D} = \frac{0.0000457 \text{ m}}{0.0525 \text{ m}} = 0.000870$$

The water mass flow rate, $\dot{m}$, is given as 26 402.39 kg/h. The velocity of the water in the pipe is
$$v = \frac{\dot{m}}{A\rho} = \frac{\left(26\,402.39 \, \frac{\text{kg}}{\text{h}}\right)\left(\frac{1 \text{ h}}{3600 \text{ s}}\right)}{(0.002165 \text{ m}^2)\left(998.2 \, \frac{\text{kg}}{\text{m}^3}\right)}$$
$$= 3.3936 \text{ m/s}$$

The Reynolds number is
$$\text{Re} = \frac{Dv\rho}{\mu} = \frac{(0.0525 \text{ m})\left(3.3936 \, \frac{\text{m}}{\text{s}}\right)\left(998.2 \, \frac{\text{kg}}{\text{m}^3}\right)}{(1 \text{ cP})\left(\frac{0.001 \, \frac{\text{kg}}{\text{m}\cdot\text{s}}}{1 \text{ cP}}\right)}$$
$$= 177\,843$$

From the Moody diagram, the friction factor, f, is given as 0.020679. The friction head loss, using the Darcy equation, is
$$h_f = \frac{fLv^2}{2Dg} = \frac{(0.020679)(50 \text{ m})\left(3.3936 \, \frac{\text{m}}{\text{s}}\right)^2}{(2)(0.0525 \text{ m})\left(9.8 \, \frac{\text{m}}{\text{s}^2}\right)}$$
$$= 11.572 \text{ m}$$

The static head, z_b, is given as 3 m. Let the datum at point a, z_a, be defined as 0 ft. Points a and b are at atmospheric pressure, so $p_a = p_b$. The general mechanical energy balance for flow from point a to point b with pump work and friction is
$$\frac{p_a}{\rho} + \left(\frac{g}{g_c}\right) z_a + \frac{v_a^2}{2g_c} + \eta W_p$$
$$= \frac{p_b}{\rho} + \left(\frac{g}{g_c}\right) z_b + \frac{v_b^2}{2g_c} + \left(\frac{g}{g_c}\right) h_f$$

The mechanical energy balance from point a to point b with no velocity change is
$$h_f = \left(\frac{g_c}{g\rho}\right)(p_a - p_b) + z_a - z_b + \left(\frac{g_c}{g}\right)\eta W_p$$
$$\left(\frac{g_c}{g}\right)\eta W_p = h_f - \left(\frac{g_c}{g\rho}\right)(p_a - p_b) - (z_a - z_b)$$
$$= 11.572 \text{ m} - 0 \text{ m} - (0 \text{ m} - 3 \text{ m})$$
$$= 14.572 \text{ m}$$
$$= h_{\text{pump}}$$

The efficiency of the pump, η, is given as 0.58. The power is
$$P = \dot{m}W_p = \frac{\dot{m}gh_{\text{pump}}}{g_c\eta}$$
$$= \frac{\left(26\,402.39 \, \frac{\text{kg}}{\text{h}}\right)\left(\frac{1 \text{ h}}{3600 \text{ s}}\right)\left(9.8 \, \frac{\text{m}}{\text{s}^2}\right)}{\left(1 \, \frac{\text{kg}\cdot\text{m}}{\text{N}\cdot\text{s}^2}\right)(0.58)}$$
$$= 1805.75 \text{ W} \quad (1800 \text{ W})$$

The answer is (C).

Why Other Options Are Wrong

(A) This incorrect answer uses a value of 58 for the efficiency instead of 0.58 when calculating the power.

(B) This incorrect answer fails to use the acceleration due to gravity when calculating the power.

(D) This incorrect answer fails to use the conversion factor from hours to seconds when calculating the power.

SOLUTION 42

The volumetric flow rate in the RS and VU lines is
$$Q_{\text{RS}} = Q_{\text{VU}} = \left(350 \, \frac{\text{gal}}{\text{min}}\right)\left(\frac{1 \text{ ft}^3}{7.48 \text{ gal}}\right)\left(\frac{1 \text{ min}}{60 \text{ sec}}\right)$$
$$= 0.7799 \text{ ft}^3/\text{sec}$$

The roughness of the pipe, ϵ, is given as 0.0002 ft. The friction coefficient in the RS line is
$$f_{\text{RS}} = \left(-2\log\frac{\epsilon}{3.7D_{\text{RS}}}\right)^{-2}$$
$$= \left(-2\log\frac{0.0002 \text{ ft}}{(3.7)(5 \text{ in})\left(\frac{1 \text{ ft}}{12 \text{ in}}\right)}\right)^{-2}$$
$$= 0.016547$$

The friction coefficients in the RS, SV, ST, TW, and WV lines are equal because the friction coefficient depends on the diameter of the line and the surface roughness, and these lines have the same diameter of 5 in.
$$f_{\text{RS}} = f_{\text{SV}} = f_{\text{ST}}$$
$$= f_{\text{TW}} = f_{\text{WV}}$$
$$= 0.016547$$

The friction coefficient in the line VU is

$$f_{VU} = \left(-2\log \frac{\epsilon}{3.7 D_{VU}}\right)^{-2}$$

$$= \left(-2\log \frac{0.0002 \text{ ft}}{(3.7)(4 \text{ in})\left(\frac{1 \text{ ft}}{12 \text{ in}}\right)}\right)^{-2}$$

$$= 0.017404$$

The head loss due to friction between points R and U, h_{RU}, is given as 67.4 ft. The gravitational acceleration, g, is 32.17 ft/sec². The Darcy equation is

$$h_f = \frac{fLv^2}{2Dg} = \frac{fL\left(\frac{Q}{A}\right)^2}{2Dg} = \frac{fLQ^2}{2DgA^2}$$

$$= \frac{fLQ^2}{2Dg\left(\frac{\pi D^2}{4}\right)^2}$$

$$= \left(\frac{8fL}{\pi^2 D^5 g}\right) Q^2$$

$$= KQ^2$$

Applying the definition of the friction head loss, by T.W. Cochran, given in the problem statement, the overall resistance coefficient is

$$K = \frac{8fL}{\pi^2 D^5 g}$$

The resistance coefficient in the RS line is

$$K_{RS} = \frac{8 f_{RS} L_{RS}}{\pi^2 D_{RS}^5 g}$$

$$= \frac{(8)(0.016547)(750 \text{ ft})}{\pi^2 \left((5 \text{ in})\left(\frac{1 \text{ ft}}{12 \text{ in}}\right)\right)^5 \left(32.17 \frac{\text{ft}}{\text{sec}^2}\right)}$$

$$= 24.8987 \text{ sec}^2/\text{ft}^5$$

The resistance coefficient in the SV line is

$$K_{SV} = \frac{8 f_{SV} L_{SV}}{\pi^2 D_{SV}^5 g}$$

$$= \frac{(8)(0.016547)(450 \text{ ft})}{\pi^2 \left((5 \text{ in})\left(\frac{1 \text{ ft}}{12 \text{ in}}\right)\right)^5 \left(32.17 \frac{\text{ft}}{\text{sec}^2}\right)}$$

$$= 14.9392 \text{ sec}^2/\text{ft}^5$$

The resistance coefficients in the lines SV, ST, WV, and TW are equal because these lines have the same equivalent length, the same diameter, and the same friction coefficient.

$$K_{SV} = K_{ST} = K_{WV} = K_{TW}$$

The resistance coefficient in the VU line is

$$K_{VU} = \frac{8 f_{VU} L_{VU}}{\pi^2 D_{VU}^5 g}$$

$$= \frac{(8)(0.017404)(750 \text{ ft})}{\pi^2 \left((4 \text{ in})\left(\frac{1 \text{ ft}}{12 \text{ in}}\right)\right)^5 \left(32.17 \frac{\text{ft}}{\text{sec}^2}\right)}$$

$$= 79.9201 \text{ sec}^2/\text{ft}^5$$

The general mechanical energy balance for flow from point R to point S with pump work and friction is

$$\frac{p_R}{\rho} + \left(\frac{g}{g_c}\right) z_R + \frac{v_R^2}{2g_c} + \eta W$$
$$= \frac{p_S}{\rho} + \left(\frac{g}{g_c}\right) z_S + \frac{v_S^2}{2g_c} + \left(\frac{g}{g_c}\right) h_f$$

The general mechanical energy balance going from point R to point S with no pump, no elevation difference, and equal velocities is

$$\frac{g_c p_R}{g \rho} = \frac{g_c p_S}{g \rho} + h_{RS} \quad \text{[I]}$$

The general mechanical energy balance going from point S to point V with no pump, no elevation difference, and equal velocities is

$$\frac{g_c p_S}{g \rho} = \frac{g_c p_V}{g \rho} + h_{SV} \quad \text{[II]}$$

The general mechanical energy balance going from point V to point U with no pump, no elevation difference, and equal velocities is

$$\frac{g_c p_V}{g \rho} = \frac{g_c p_U}{g \rho} + h_{VU} \quad \text{[III]}$$

Adding equations I, II, and III and then simplifying gives

$$\frac{g_c p_R}{g \rho} = \frac{g_c p_U}{g \rho} + h_{RS} + h_{SV} + h_{VU}$$

Solving for the pressure drop head from R to U gives

$$\frac{g_c (p_R - p_U)}{g \rho} = h_{RU}$$

$$= h_{RS} + h_{SV} + h_{VU}$$

Rearranging,

$$h_{RU} - h_{RS} - h_{VU} = h_{SV}$$

Applying the T.W. Cochran relationship gives

$$h_{\mathrm{RU}} - K_{\mathrm{RS}}Q_{\mathrm{RS}}^2 - K_{\mathrm{VU}}Q_{\mathrm{VU}}^2 = K_{\mathrm{SV}}Q_{\mathrm{SV}}^2$$

$$Q_{\mathrm{SV}} = \sqrt{\frac{h_{\mathrm{RU}} - K_{\mathrm{RS}}Q_{\mathrm{RS}}^2 - K_{\mathrm{VU}}Q_{\mathrm{VU}}^2}{K_{\mathrm{SV}}}}$$

$$= \sqrt{\frac{67.4\text{ ft} - \left(24.8987\,\frac{\sec^2}{\text{ft}^5}\right)\left(0.7799\,\frac{\text{ft}^3}{\sec}\right)^2 - \left(79.9201\,\frac{\sec^2}{\text{ft}^5}\right)\left(0.7799\,\frac{\text{ft}^3}{\sec}\right)^2}{14.9392\,\frac{\sec^2}{\text{ft}^5}}}$$

$$= \left(0.49393\,\frac{\text{ft}^3}{\sec}\right)\left(7.48\,\frac{\text{gal}}{\text{ft}^3}\right)\left(60\,\frac{\sec}{\min}\right)$$

$$= 221.67\text{ gal/min} \quad (220\text{ gal/min})$$

The answer is (D).

Why Other Options Are Wrong

(A) This incorrect answer is the volumetric flow rate in the SV line in cubic feet per second.

(B) This incorrect answer does not multiply by 60 sec/min when calculating the flow rate in the SV line.

(C) This incorrect answer does not multiply by 7.48 gal/ft^3 when calculating the flow rate in the SV line.

SOLUTION 43

The density of the fluid, ρ, is given as 56.6 lbm/ft^3. The viscosity of the fluid, μ, is given as 0.00027 lbm/sec-ft. The volumetric flow rate of the fluid at branch AB is

$$Q_{\mathrm{AB}} = \left(400\,\frac{\text{gal}}{\min}\right)\left(\frac{1\text{ min}}{60\text{ sec}}\right)\left(\frac{1\text{ ft}^3}{7.48\text{ gal}}\right)$$

$$= 0.8913\text{ ft}^3/\sec$$

The flow rates of the fluid at branches AB and EF are equal to the total flow rate in the pipe network.

$$Q_{\mathrm{AB}} = Q_{\mathrm{EF}} = Q_t$$
$$= 0.8913\text{ ft}^3/\sec$$

The roughness of the pipe, ϵ, is given as 0.0002 ft. The friction factor in the AB line is

$$f_{\mathrm{AB}} = \left(-2\log\frac{\epsilon}{3.7D_{\mathrm{AB}}}\right)^{-2}$$

$$= \left(-2\log\frac{0.0002\text{ ft}}{(3.7)(0.41667\text{ ft})}\right)^{-2}$$

$$= 0.016547$$

The friction factors in the AB, EF, BC, FG, and CG lines are equal because these lines have the same diameter and surface roughness.

$$f_{\mathrm{AB}} = f_{\mathrm{EF}} = f_{\mathrm{BC}}$$
$$= f_{\mathrm{FG}} = f_{\mathrm{CG}}$$
$$= 0.016547$$

The friction factor in the BF line is

$$f_{\mathrm{BF}} = \left(-2\log\frac{\epsilon}{3.7D_{\mathrm{BF}}}\right)^{-2}$$

$$= \left(-2\log\frac{0.0002\text{ ft}}{(3.7)(0.33333\text{ ft})}\right)^{-2}$$

$$= 0.017404$$

The head loss due to friction between points A and E, h_{AE}, is given as 49.79 ft. The gravitational acceleration, g, is 32.17 ft/sec^2. The average velocity of the fluid in the pipe is

$$\mathrm{v} = \frac{Q}{A}$$

From the Darcy equation, the head loss due to friction is

$$h_f = \frac{fL\mathrm{v}^2}{2Dg} = \frac{fL\left(\dfrac{Q}{A}\right)^2}{2Dg}$$

$$= \frac{fLQ^2}{2Dg\left(\dfrac{\pi D^2}{4}\right)^2} = \left(\frac{8fL}{\pi^2 D^5 g}\right)Q^2$$

$$= KQ^2$$

The overall resistance coefficient is

$$K = \frac{8fL_e}{\pi^2 D^5 g}$$

The resistance coefficient in the AB line is

$$K_{\mathrm{AB}} = \frac{8f_{\mathrm{AB}}L_{e\mathrm{AB}}}{\pi^2 D_{\mathrm{AB}}^5 g}$$

$$= \frac{(8)(0.016547)(800\text{ ft})}{\pi^2(0.41667\text{ ft})^5\left(32.17\,\dfrac{\text{ft}}{\sec^2}\right)}$$

$$= 26.5575\text{ sec}^2/\text{ft}^5$$

The resistance coefficient in the EF line equals the resistance coefficient in the AB line because these two lines have the same equivalent length and the same friction factor. The resistance coefficient in the EF line, K_{EF}, is 26.5575 sec^2/ft^5.

The resistance coefficient in the BF line is

$$K_{BF} = \frac{8 f_{BF} L_{eBF}}{\pi^2 D_{BF}^5 g}$$

$$= \frac{(8)(0.017404)(400 \text{ ft})}{\pi^2 (0.33333 \text{ ft})^5 \left(32.17 \frac{\text{ft}}{\text{sec}^2}\right)}$$

$$= 42.6261 \text{ sec}^2/\text{ft}^5$$

Applying the Bernoulli equation around ABFE gives

$$h_{AE} - h_{AB} - h_{FE} = h_{BF}$$

Applying the T.W. Cochran relationship gives

$$h_{AE} - K_{AB} Q_{AB}^2 - K_{EF} Q_{EF}^2 = K_{BF} Q_{BF}^2$$

The resistance coefficient in the AB line equals the resistance coefficient in the EF line because these two lines have the same equivalent length and the same friction factor. Also, the flow rate in line AB equals the volumetric flow rate in line EF because lines AB and EF are terminal lines.

$$K_{AB} Q_{AB}^2 = K_{EF} Q_{EF}^2$$

Replacing and simplifying gives

$$h_{AE} - 2 K_{AB} Q_{AB}^2 = K_{BF} Q_{BF}^2$$

$$Q_{BF} = \sqrt{\frac{h_{AE} - 2 K_{AB} Q_{AB}^2}{K_{BF}}}$$

$$= \sqrt{\frac{49.79 \text{ ft} - (2)\left(26.5575 \frac{\text{sec}^2}{\text{ft}^5}\right)\left(0.8913 \frac{\text{ft}^3}{\text{sec}}\right)^2}{42.6261 \frac{\text{sec}^2}{\text{ft}^5}}}$$

$$= 0.4221 \text{ ft}^3/\text{sec}$$

The Reynolds number in the BF line is

$$\text{Re}_{BF} = \frac{D v \rho}{\mu} = \frac{4 \rho Q_{BF}}{\pi D_{BF} \mu}$$

$$= \frac{(4)\left(56.6 \frac{\text{lbm}}{\text{ft}^3}\right)\left(0.4221 \frac{\text{ft}^3}{\text{sec}}\right)}{\pi (0.33333 \text{ ft}) \left(0.00027 \frac{\text{lbm}}{\text{sec-ft}}\right)}$$

$$= 337{,}989.91 \quad (338{,}000)$$

The answer is (C).

Why Other Options Are Wrong

(A) This incorrect answer is the Reynolds number in the DH line.

(B) This incorrect answer is the Reynolds number in the BC line.

(D) This incorrect answer is the Reynolds number in the AB line.

SOLUTION 44

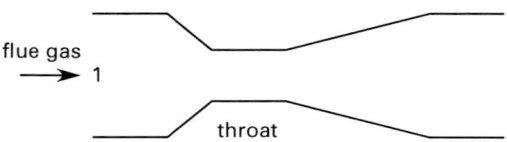

The ratio of specific heat capacities, k, is given as 1.41. The molecular weight of CO_2, MW_{CO_2}, is given as 44 lbm/lbmol. The molecular weight of the flue gas, MW_{fg}, is given as 29.3 lbm/lbmol. The mole fraction of CO_2 in the flue gas, x_{CO_2}, is given as 0.5%. The diameter of the pipe, d_1, is given as 3.068 in. The diameter of the throat, d, is given as 1.25 in.

The area of the throat of the venturi is

$$A = \frac{\pi d^2}{4} = \frac{\pi \left((1.25 \text{ in})\left(\frac{1 \text{ ft}}{12 \text{ in}}\right)\right)^2}{4}$$

$$= 0.00852 \text{ ft}^2$$

The ratio of throat diameter to pipe diameter is

$$\beta = \frac{d}{d_1} = \frac{1.25 \text{ in}}{3.068 \text{ in}}$$

$$= 0.407$$

The gravitational constant, g_c, is 32.17 lbm-ft/lbf-sec^2. The manometer reading is given as

$$\Delta p = (20 \text{ in Hg}) \left(0.491 \frac{\frac{\text{lbf}}{\text{in}^2}}{\text{in Hg}}\right)$$

$$= 9.82 \text{ lbf/in}^2$$

The pressure before the venturi meter, p_a, is given as 14.5 lbf/in^2. The pressure at the throat is

$$p_b = p_a - \Delta p = 14.5 \frac{\text{lbf}}{\text{in}^2} - 9.82 \frac{\text{lbf}}{\text{in}^2}$$

$$= 4.68 \text{ lbf/in}^2$$

The discharge coefficient is given as a straight line having a slope of 0.0123 and an intercept of 0.9878. The discharge coefficient is

$$C = 0.9878 + 0.0123 \beta$$

$$= 0.9878 + (0.0123)(0.407)$$

$$= 0.9928$$

The pressure ratio between the pressure before the venturi meter and the pressure at the throat is defined as

$$\phi = \frac{p_b}{p_a} = \frac{4.68 \ \frac{\text{lbf}}{\text{in}^2}}{14.5 \ \frac{\text{lbf}}{\text{in}^2}} = 0.3228$$

For isentropic flow through a venturi meter, the expansion factor ratio is

$$Y = \sqrt{\frac{\left(\frac{k}{k-1}\right) \phi^{2/k} \left(1 - \phi^{\frac{k-1}{k}}\right)(1-\beta^4)}{(1-\phi)(1-\beta^4 \phi^{2/k})}}$$

$$= \sqrt{\frac{\left(\frac{1.41}{1.41-1}\right)(0.3228)^{2/1.41}\left(1-(0.3228)^{\frac{1.41-1}{1.41}}\right) \times \left(1-(0.407)^4\right)}{(1-0.3228)\left(1-(0.407)^4(0.3228)^{2/1.41}\right)}}$$

$$= 0.52904$$

The temperature of the flue gas feed is

$$T_1 = 320°\text{F} + 460° = 780°\text{R}$$

The universal gas constant, R, is 10.73 lbf-ft^3/in^2-lbmol-°R. Because the flue gas is considered an ideal gas, the density of the flue gas at the throat is calculated at standard conditions.

$$\rho = \frac{p_a(\text{MW}_\text{fg})}{RT_1} = \frac{\left(14.5 \ \frac{\text{lbf}}{\text{in}^2}\right)\left(29.3 \ \frac{\text{lbm}}{\text{lbmol}}\right)}{\left(10.73 \ \frac{\text{lbf-ft}^3}{\text{lbmol-in}^2\text{-°R}}\right)(780°\text{R})}$$

$$= 0.05076 \ \text{lbm/ft}^3$$

The volumetric flow rate of the flue gas is

$$Q = YCA\sqrt{\frac{2g_c(p_a - p_b)}{\rho}}$$

$$= (0.52904)(0.9928)(0.00852 \ \text{ft}^2)$$

$$\times \sqrt{\frac{(2)\left(32.17 \ \frac{\text{lbm-ft}}{\text{lbf-sec}^2}\right)\left(14.5 \ \frac{\text{lbf}}{\text{in}^2} - 4.68 \ \frac{\text{lbf}}{\text{in}^2}\right) \times \left(144 \ \frac{\text{in}^2}{\text{ft}^2}\right)}{0.05076 \ \frac{\text{lbm}}{\text{ft}^3}}}$$

$$= 5.8070 \ \text{ft}^3/\text{sec}$$

The mass flow rate of CO_2 is

$$\dot{m}_{CO_2} = \frac{Q\rho(\text{MW}_{CO_2})x_{CO_2}}{\text{MW}_\text{fg}}$$

$$= \frac{\left(5.8070 \ \frac{\text{ft}^3}{\text{sec}}\right)\left(0.05076 \ \frac{\text{lbm}}{\text{ft}^3}\right)\left(3600 \ \frac{\text{sec}}{\text{hr}}\right) \times \left(44 \ \frac{\text{lbm}}{\text{lbmol}}\right)(0.005)}{29.3 \ \frac{\text{lbm}}{\text{lbmol}}}$$

$$= 7.9677 \ \text{lbm/hr} \quad (8 \ \text{lbm/hr})$$

The answer is (B).

Why Other Options Are Wrong

(A) This incorrect answer omits the conversion factor from seconds to hour when calculating the mass flow rate.

(C) This incorrect answer uses the area of the pipe instead of the area of the throat when calculating the mass flow rate.

(D) This incorrect answer uses an expansion coefficient equal to one when calculating the volumetric flow rate of the flue gas.

SOLUTION 45

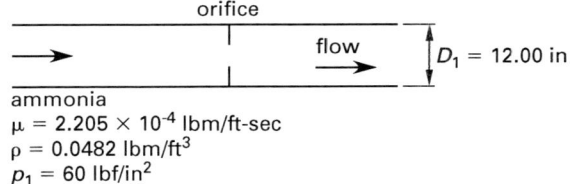

The area of the pipe, A, is given as 0.7854 ft^2. The diameter of the pipe is

$$D_1 = (12.00 \ \text{in})\left(\frac{1 \ \text{ft}}{12 \ \text{in}}\right) = 1.00 \ \text{ft}$$

The ammonia volumetric flow rate is given as

$$Q = \left(6000 \ \frac{\text{ft}^3}{\text{min}}\right)\left(\frac{1 \ \text{min}}{60 \ \text{sec}}\right) = 100 \ \text{ft}^3/\text{sec}$$

The velocity of the ammonia in the pipe is

$$v = \frac{Q}{A} = \frac{100 \ \frac{\text{ft}^3}{\text{sec}}}{0.7854 \ \text{ft}^2}$$

$$= 127.32 \ \text{ft/sec}$$

The Reynolds number is

$$\text{Re} = \frac{D_1 v \rho}{\mu} = \frac{(1.00 \text{ ft})\left(127.32 \frac{\text{ft}}{\text{sec}}\right)\left(0.0482 \frac{\text{lbm}}{\text{ft}^3}\right)}{2.205 \times 10^{-4} \frac{\text{lbm}}{\text{ft-sec}}}$$

$$= 27{,}831$$

Because the Reynolds number is larger than 20,000, the flow is in a fully turbulent region. For Reynolds numbers larger than 20,000, the instrument pressure is

$$\Delta p_{\text{ins}} = (127 \text{ in H}_2\text{O})\left(0.036092 \frac{\frac{\text{lbf}}{\text{in}^2}}{\text{in H}_2\text{O}}\right)$$

$$= 4.584 \text{ lbf/in}^2$$

The pressure after the orifice is

$$p_2 = p_1 - \Delta p_{\text{ins}} = 60 \frac{\text{lbf}}{\text{in}^2} - 4.58 \frac{\text{lbf}}{\text{in}^2}$$

$$= 55.42 \text{ lbf/in}^2$$

The ratio of the pressure after the orifice to the pressure before the orifice is

$$\frac{p_2}{p_1} = \frac{55.42 \frac{\text{lbf}}{\text{in}^2}}{60 \frac{\text{lbf}}{\text{in}^2}} = 0.924$$

This ratio must be greater than 0.53, which is the critical pressure ratio at which airflow becomes sonic. The ratio of the diameter of the orifice to the internal diameter of the pipe is

$$\beta = \frac{d_o}{d_1}$$

The diameter of the orifice is

$$d_o = d_1 \beta = (12.00 \text{ in})\beta$$

The ratio of the specific heats, k, is given as 1.3. The square-edged orifice coefficient, C_d, is given as 0.595. For Reynolds numbers larger than 20,000, the flow coefficient is

$$C_f = \frac{C_d}{\sqrt{1 - \left(\frac{d_o}{d_1}\right)^4}} = \frac{0.595}{\sqrt{1 - \beta^4}} \quad [\text{I}]$$

The expansion coefficient is

$$Y = 1 - \left(\frac{0.41 + 0.35\beta^4}{k}\right)\left(1 - \frac{p_2}{p_1}\right)$$

$$= 1 - \left(\frac{0.41 + 0.35\beta^4}{k}\right)(1 - 0.924)$$

$$= 1 - \left(\frac{0.41 + 0.35\beta^4}{1.3}\right)(0.076) \quad [\text{II}]$$

The gravitational constant, g_c, is 32.17 lbm-ft/lbf-sec^2. The volumetric flow rate is

$$Q = \tfrac{\pi}{4} d_o^2 C_f Y \sqrt{\frac{2 g_c \Delta p_{\text{ins}}}{\rho}}$$

$$= \tfrac{\pi}{4} d_o^2 \left(\frac{1 \text{ ft}^2}{144 \text{ in}^2}\right) C_f Y$$

$$\times \sqrt{\frac{(2)\left(32.17 \frac{\text{lbm-ft}}{\text{lbf-sec}^2}\right)\left(4.584 \frac{\text{lbf}}{\text{in}^2}\right) \times \left(144 \frac{\text{in}^2}{\text{ft}^2}\right)}{0.0482 \frac{\text{lbm}}{\text{ft}^3}}}$$

In the preceding equation, the flow and expansion coefficients are expressed in terms of the diameter of the orifice. Substituting yields

$$100 \frac{\text{ft}^3}{\text{sec}} = \tfrac{\pi}{4} d_o^2 \left(\frac{1 \text{ ft}^2}{144 \text{ in}^2}\right) \left(\frac{0.595}{\sqrt{1 - \left(\frac{d_o}{12.00 \text{ in}}\right)^4}}\right)$$

$$\times \left(1 - \left(\frac{0.41 + (0.35) \times \left(\frac{d_o}{12.00 \text{ in}}\right)^4}{1.3}\right)(0.076)\right)$$

$$\times \sqrt{\frac{(2)\left(32.17 \frac{\text{lbm-ft}}{\text{lbf-sec}^2}\right)\left(4.584 \frac{\text{lbf}}{\text{in}^2}\right) \times \left(144 \frac{\text{in}^2}{\text{ft}^2}\right)}{0.0482 \frac{\text{lbm}}{\text{ft}^3}}}$$

The value of the diameter of the orifice can be determined by solving the preceding equation numerically for d_o using a programmable calculator.

$$d_o = 5.73 \text{ in} \quad (5.7 \text{ in})$$

The answer is (C).

Alternate Solution

The preceding equation could also be solved iteratively by giving values to d_o. The value of d_o that makes Q equal to 100 ft^3/sec is the solution.

Q (ft^3/sec)	d_o (in)	Eq. I	Eq. II
181	7.0	1.7486	0.413586
155	6.5	2.0279	0.352311
115	6	2.3800	0.262615
100	5.85	2.5035	0.227749

Why Other Options Are Wrong

(A) This incorrect answer omits the conversion factor from square feet to square inches when calculating the diameter.

(B) This incorrect answer is the diameter calculated with the flow and the expansion coefficients each equal to 1.

(D) This incorrect answer is calculated by multiplying the internal diameter of the pipe by the square-edged orifice coefficient.

HEAT TRANSFER
SOLUTION 46

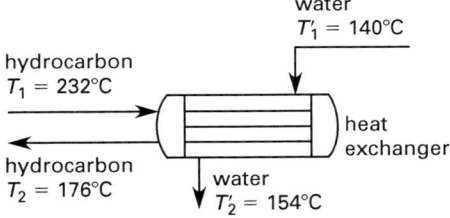

The mass flow rate of the hydrocarbon, $\dot{m}_h$, is given as 1000 kg/h. The specific heat capacity of the hydrocarbon, c_p, is given as 2.594 kJ/kg·°C. The specific heat capacity of the water, $c_{p,\text{H}_2\text{O}}$, is given as 2.427 kJ/kg·°C. The water mass flow rate in the heat exchanger is

$$\dot{m}_w = \frac{\dot{m}_h c_p (T_1 - T_2)}{c_{p,\text{H}_2\text{O}} (T'_2 - T'_1)}$$

$$= \frac{\left(1000 \, \frac{\text{kg}}{\text{h}}\right)\left(2.594 \, \frac{\text{kJ}}{\text{kg·°C}}\right)(232°\text{C} - 176°\text{C})}{\left(2.427 \, \frac{\text{kJ}}{\text{kg·°C}}\right)(154°\text{C} - 140°\text{C})}$$

$$= 4275 \text{ kg/h} \quad (4300 \text{ kg/h})$$

The answer is (D).

Why Other Options Are Wrong

(A) This answer erroneously exchanges the temperatures of the cold fluid with those of the hot fluid.

(B) This answer erroneously exchanges the exit temperature of the hydrocarbon mixture with that of the water.

(C) This answer erroneously exchanges the heat capacities of the hydrocarbon mixture with those of the water.

SOLUTION 47

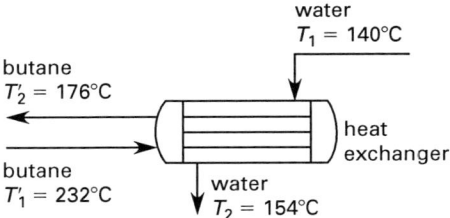

The overall heat-transfer coefficient based on the outside area, U_o, is given as 205 kJ/m²·h·°C. The logarithmic mean temperature difference (LMTD) is

$$\Delta T_{\text{lm}} = \frac{(T'_1 - T_2) - (T'_2 - T_1)}{\ln \dfrac{T'_1 - T_2}{T'_2 - T_1}}$$

$$= \frac{(232°\text{C} - 154°\text{C}) - (176°\text{C} - 140°\text{C})}{\ln \dfrac{232°\text{C} - 154°\text{C}}{176°\text{C} - 140°\text{C}}}$$

$$= 54.32°\text{C}$$

Calculating the R and S parameters to use in looking up the LMTD correction factor correlation,

$$R = \frac{T_1 - T_2}{T'_2 - T'_1} = \frac{140°\text{C} - 154°\text{C}}{176°\text{C} - 232°\text{C}}$$

$$= 0.25$$

$$S = \frac{T'_2 - T'_1}{T_1 - T'_1} = \frac{176°\text{C} - 232°\text{C}}{140°\text{C} - 232°\text{C}}$$

$$= 0.609$$

With these two values and from the F_T chart, the correction factor for the one-shell-pass/two-tube-passes heat exchanger, F_T, is 0.96. Using the correction factor, the corrected temperature is

$$(\Delta T_{\text{lm}})_c = F_T \Delta T_{\text{lm}} = (0.96)(54.32°\text{C})$$

$$= 52.15°\text{C}$$

The specific heat capacity, c_p, is given as 2.594 kJ/kg·°C. The rate of heat exchanged is

$$Q = \dot{m}_h c_p (T'_1 - T'_2)$$

$$= \left(1000 \, \frac{\text{kg}}{\text{h}}\right)\left(2.594 \, \frac{\text{kJ}}{\text{kg·°C}}\right)(232°\text{C} - 176°\text{C})$$

$$= 145\,264 \text{ kJ/h}$$

This rate is the same as

$$Q = U_o A_o F_T \Delta T_{\text{lm}}$$

$$= U_o A_o (\Delta T_{\text{lm}})_c$$

Solving for the area of the heat exchanger gives

$$A_o = \frac{Q}{U_o (\Delta T_{lm})_c}$$

$$= \frac{145\,264 \ \frac{\text{kJ}}{\text{h}}}{\left(205 \ \frac{\text{kJ}}{\text{m}^2 \cdot \text{h} \cdot °\text{C}}\right)(52.15°\text{C})}$$

$$= 13.59 \ \text{m}^2 \quad (13.6 \ \text{m}^2)$$

The answer is (D).

Why Other Options Are Wrong

(A) This incorrect answer calculates the area using the overall heat-transfer coefficient based on the internal area instead of the heat-transfer coefficient based on the outside area.

(B) This incorrect answer calculates the rate of heat transfer using the heat capacity of water instead of the heat capacity of butane.

(C) This incorrect answer does not correct the LMTD.

SOLUTION 48

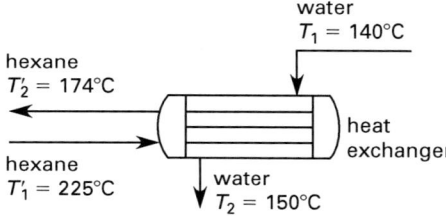

The overall heat-transfer coefficient based on the outside area, U_o, is given as 800 Btu/ft²-hr-°F. The heat capacity of the hexanes mixture, c_p, is given as 0.82 Btu/lbm-°F. The temperature difference between the hexanes mixture at the entrance of the heat exchanger and at the exit is

$$\Delta T = T'_1 - T'_2 = 225°\text{F} - 174°\text{F}$$
$$= 51°\text{F}$$

The rate of heat exchange is

$$Q = \dot{m}_h c_p \Delta T$$
$$= \left(50{,}000 \ \frac{\text{lbm}}{\text{hr}}\right)\left(0.82 \ \frac{\text{Btu}}{\text{lbm-}°\text{F}}\right)(51°\text{F})$$
$$= 2{,}091{,}000 \ \text{Btu/hr}$$

The logarithmic mean temperature difference (LMTD) is

$$\Delta T_{lm} = \frac{(T'_1 - T_2) - (T'_2 - T_1)}{\ln \frac{T'_1 - T_2}{T'_2 - T_1}}$$

$$= \frac{(225°\text{F} - 150°\text{F}) - (174°\text{F} - 140°\text{F})}{\ln \frac{225°\text{F} - 150°\text{F}}{174°\text{F} - 140°\text{F}}}$$

$$= 51.82°\text{F}$$

The R and S parameters are needed to look up the LMTD correction factor correlation.

$$R = \frac{T_1 - T_2}{T'_2 - T'_1} = \frac{140°\text{F} - 150°\text{F}}{174°\text{F} - 225°\text{F}}$$
$$= 0.196$$

$$S = \frac{T'_2 - T'_1}{T_1 - T'_1} = \frac{174°\text{F} - 225°\text{F}}{140°\text{F} - 225°\text{F}}$$
$$= 0.60$$

With these values for R and S, from the F_T chart, the correction factor for the one-shell-pass/two-tube-passes heat exchanger, F_T, is 0.97. Using this correction factor, the corrected temperature is

$$(\Delta T_{lm})_c = F_T \Delta T_{lm} = (0.97)(51.82°\text{F}) = 50.27°\text{F}$$

The area of the heat exchanger is

$$A_o = \frac{Q}{U_o(\Delta T_{lm})_c} = \frac{2{,}091{,}000 \ \frac{\text{Btu}}{\text{hr}}}{\left(800 \ \frac{\text{Btu}}{\text{ft}^2\text{-hr-}°\text{F}}\right)(50.27°\text{F})}$$

$$= 51.99 \ \text{ft}^2$$

The tube diameter, D, is given as 0.277 ft. The number of tubes, n, is given as 10. The area of the heat exchanger, A_o, is $\pi D L n$. The length of the tubes is

$$L = \frac{A_o}{\pi D n} = \frac{51.99 \ \text{ft}^2}{\pi (0.277 \ \text{ft})(10)}$$
$$= 5.97 \ \text{ft} \quad (6.0 \ \text{ft})$$

The answer is (B).

Why Other Options Are Wrong

(A) This answer does not correct the LMTD and, therefore, uses an uncorrected temperature when calculating the rate of heat exchanged.

(C) This incorrect answer uses the heat capacity of the water instead of the heat capacity of the hexanes mixture when calculating the rate of heat transferred from the hydrocarbon to the water.

(D) This incorrect answer uses the overall heat-transfer coefficient based on the inside area instead of the heat-transfer coefficient based on the outside area.

SOLUTION 49

The mass flow rate of process stream C1, $\dot{m}_1$, is given as 1 kg/min. The mass flow rate of process stream C2, $\dot{m}_2$, is given as 2 kg/min. The mass flow rate of process stream C3 is

$$\dot{m}_3 = \dot{m}_1 + \dot{m}_2 = 1\ \frac{\text{kg}}{\text{min}} + 2\ \frac{\text{kg}}{\text{min}}$$
$$= 3\ \text{kg/min}$$

An energy balance around heat exchanger 1 gives the caloric energy gain by process stream C1 in heat exchanger 1.

$$q_1 = \dot{m}_1 c_p (T_{1,\text{out}} - T_{1,\text{in}})$$
$$= \left(1\ \frac{\text{kg}}{\text{min}}\right)\left(60\ \frac{\text{kJ}}{\text{kg·K}}\right)(480\text{K} - 300\text{K})\left(\frac{1\ \text{min}}{60\ \text{s}}\right)$$
$$= 180\ \text{kJ/s}\quad (180\ \text{kW})$$

The caloric energy gain by process stream C2 in heat exchanger 2 is

$$q_2 = \dot{m}_2 c_p (T_{2,\text{out}} - T_{2,\text{in}})$$
$$= \left(2\ \frac{\text{kg}}{\text{min}}\right)\left(60\ \frac{\text{kJ}}{\text{kg·K}}\right)(480\text{K} - 300\text{K})\left(\frac{1\ \text{min}}{60\ \text{s}}\right)$$
$$= 360\ \text{kJ/s}\quad (360\ \text{kW})$$

An energy balance around heat exchanger 3 yields the caloric energy given by the reactor effluent in heat exchanger 3. The caloric energy is

$$q_3 = \dot{m}_3 c_p (T_{3,\text{in}} - T_{3,\text{out}})$$
$$= \left(3\ \frac{\text{kg}}{\text{min}}\right)\left(60\ \frac{\text{kJ}}{\text{kg·K}}\right)(500\text{K} - 400\text{K})\left(\frac{1\ \text{min}}{60\ \text{s}}\right)$$
$$= 300\ \text{kJ/s}\quad (300\ \text{kW})$$

The heating area of heat exchanger 1 is

$$A_1 = \frac{q_1}{U_1(T_{1,\text{out}} - T_{1,\text{in}})}$$
$$= \left(\frac{180\ \text{kW}}{\left(800\ \frac{\text{W}}{\text{m}^2\text{·K}}\right)(480\text{K} - 300\text{K})}\right)\left(1000\ \frac{\text{W}}{\text{kW}}\right)$$
$$= 1.25\ \text{m}^2$$

The heating area of heat exchanger 2 is

$$A_2 = \frac{q_2}{U_2(T_{2,\text{out}} - T_{2,\text{in}})}$$
$$= \left(\frac{360\ \text{kW}}{\left(1200\ \frac{\text{W}}{\text{m}^2\text{·K}}\right)(480\text{K} - 300\text{K})}\right)\left(1000\ \frac{\text{W}}{\text{kW}}\right)$$
$$= 1.67\ \text{m}^2$$

The heating area of heat exchanger 3 is

$$A_3 = \frac{q_3}{U_3(T_{3,\text{in}} - T_{3,\text{out}})}$$
$$= \left(\frac{300\ \text{kW}}{\left(800\ \frac{\text{W}}{\text{m}^2\text{·K}}\right)(500\text{K} - 400\text{K})}\right)\left(1000\ \frac{\text{W}}{\text{kW}}\right)$$
$$= 3.75\ \text{m}^2$$

The heat exchanger capital cost per unit area, $C_{c,\text{unit}}$, is given as \$2174/m². The total heat exchanger capital cost is

$$C_c = C_{c,\text{unit}}(A_1 + A_2 + A_3)$$
$$= \left(\frac{\$2714}{\text{m}^2}\right)(1.25\ \text{m}^2 + 1.67\ \text{m}^2 + 3.75\ \text{m}^2)$$
$$= \$18{,}102$$

The heat exchanger annualized investment is

$$H_{\text{inv}} = (\text{factor})C_c$$
$$= (0.2\ \text{yr}^{-1})(\$18{,}102)$$
$$= \$3620/\text{yr}$$

The hot water utility annual cost is

$$C_{\text{hw}} = (\text{HD}_{\text{hw}})\,C_{u,\text{hw}}$$
$$= (450\ \text{kW})\left(\frac{\$8 \times 10^{-7}}{\text{kJ}}\right)\left(1\ \frac{\text{kJ}}{\text{kW·s}}\right)$$
$$\quad \times \left(8000\ \frac{\text{h}}{\text{yr}}\right)\left(3600\ \frac{\text{s}}{\text{h}}\right)$$
$$= \$10{,}368/\text{yr}$$

The steam utility annual cost is

$$C_S = (\text{HD}_S)\,C_{u,S}$$
$$= (540\ \text{kW})\left(\frac{\$5 \times 10^{-6}}{\text{kJ}}\right)\left(1\ \frac{\text{kJ}}{\text{kW·s}}\right)$$
$$\quad \times \left(8000\ \frac{\text{h}}{\text{yr}}\right)\left(3600\ \frac{\text{s}}{\text{h}}\right)$$
$$= \$77{,}760/\text{yr}$$

The cooling water utility annual cost is

$$C_{cw} = (HD_{cw})C_{u,cw}$$
$$= (300 \text{ kW})\left(\frac{\$3 \times 10^{-7}}{\text{kJ}}\right)\left(1\ \frac{\text{kJ}}{\text{kW·s}}\right)$$
$$\times \left(8000\ \frac{\text{h}}{\text{yr}}\right)\left(3600\ \frac{\text{s}}{\text{h}}\right)$$
$$= \$2592/\text{yr}$$

The annual capital and utility cost is

$$C_{c,u} = H_{inv} + C_{hw} + C_S + C_{cw}$$
$$= \frac{\$3620}{\text{yr}} + \frac{\$10{,}368}{\text{yr}} + \frac{\$77{,}760}{\text{yr}} + \frac{\$2592}{\text{yr}}$$
$$= \$94{,}340/\text{yr} \quad (\$94{,}000/\text{yr})$$

The answer is (D).

Why Other Options Are Wrong

(A) This incorrect answer omits the hot water cost when calculating the annual capital and utility cost.

(B) This incorrect answer omits the heat exchanger investment when calculating the annual capital and utility cost.

(C) This incorrect answer omits the cooling water cost when calculating the annual capital and utility cost.

SOLUTION 50

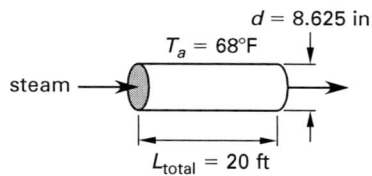

The combined heat-transfer coefficient, h_a, is given as 2.4 Btu/ft^2-hr-°F. The external diameter of the pipe is given as

$$D = (8.625 \text{ in})\left(\frac{1 \text{ ft}}{12 \text{ in}}\right)$$
$$= 0.719 \text{ ft}$$

Note that the rate of heat loss is given per foot of pipe length. Make the calculations in terms of a 1 ft length of pipe. The external area of the pipe per foot of pipe length is

$$A = \pi D L$$
$$= \pi (0.719 \text{ ft})(1 \text{ ft})$$
$$= 2.26 \text{ ft}^2$$

The rate of heat loss per foot of pipe length is given as

$$Q = 2115.5 \text{ Btu/hr}$$
$$= h_a A (T_s - T_a)$$

Solving for the temperature on the surface of the pipe gives

$$T_s = \frac{Q}{h_a A} + T_a$$
$$= \frac{2115.5\ \frac{\text{Btu}}{\text{hr}}}{\left(2.4\ \frac{\text{Btu}}{\text{ft}^2\text{-hr-°F}}\right)(2.26 \text{ ft}^2)} + 68°\text{F}$$
$$= 458.03°\text{F} \quad (460°\text{F})$$

The answer is (D).

Why Other Options Are Wrong

(A) This incorrect answer ignores the pipe-to-air resistance. This answer assumes that because the surface is in contact with air at 68°F, the pipe is also at the same temperature.

(B) In this incorrect answer, the full length of pipe is considered instead of one-foot of pipe when calculating the surface of heat transfer.

(C) In this incorrect answer, the ambient temperature is subtracted instead of added in the calculation of the surface temperature.

SOLUTION 51

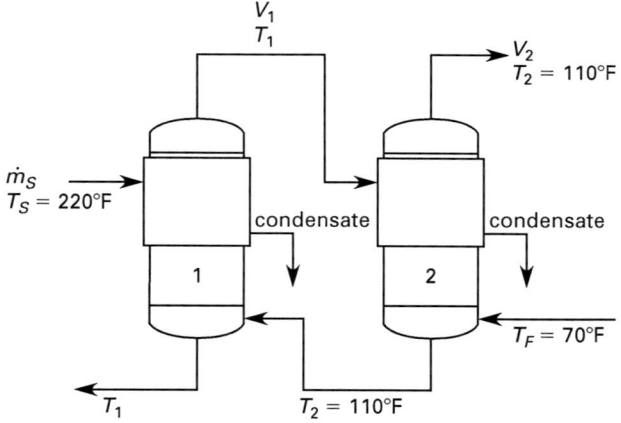

The overall heat-transfer coefficient in the first evaporator, U_1, is given as 450 Btu/ft^2-hr-°F. The overall heat-transfer coefficient in the second evaporator, U_2, is given as 315 Btu/ft^2-hr-°F. The rate of heat transferred in the first evaporator is approximately equal to the rate of heat transferred in the second evaporator.

$$Q_1 = A_1 U_1 \Delta T_1 \approx Q_2 = A_2 U_2 \Delta T_2$$

In this problem,
$$T_1 = 2T_2$$

The area of the first evaporator is equal to the area of the second evaporator.
$$A_1 = A_2 = A$$

The temperature change in the first effect is
$$\Delta T_1 = T_S - T_1$$

The temperature change in the second effect is
$$\Delta T_2 = T_1 - T_2$$
$$AU_1 \Delta T_1 = AU_2 \Delta T_2$$

Substituting for ΔT_1 and ΔT_2 and solving for T_1 gives

$$T_1 = T_2 \left(\frac{2U_1 + U_2}{U_1 + U_2} \right)$$
$$= (110°F) \left(\frac{(2)\left(450 \frac{\text{Btu}}{\text{ft}^2\text{-hr-}°F}\right) + 315 \frac{\text{Btu}}{\text{ft}^2\text{-hr-}°F}}{450 \frac{\text{Btu}}{\text{ft}^2\text{-hr-}°F} + 315 \frac{\text{Btu}}{\text{ft}^2\text{-hr-}°F}} \right)$$
$$= 174.7°F \quad (170°F)$$

The answer is (C).

Why Other Options Are Wrong

(A) This answer erroneously assumes that, because the temperature of the condenser in the second evaporator is 110°F, this must also be the temperature of the vapors produced in the first evaporator.

(B) This incorrect answer exchanges the values of the overall heat-transfer coefficients.

(D) This incorrect answer is the temperature of the saturated steam. This answer assumes that the first evaporator reaches the temperature of the saturated steam entering the system.

SOLUTION 52

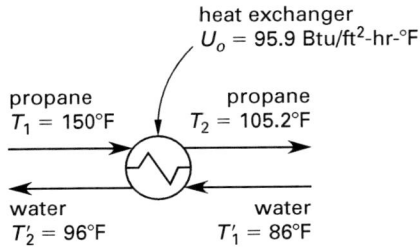

The heat capacity of propane, c_p, is given as 0.39 Btu/lbm-°F. The latent heat of propane, λ, is given as 138.1 Btu/lbm. The mass flow rate of propane in the heat exchanger, $\dot{m}$, is given as 2.7×10^6 lbm/hr. The heat transferred from the propane to the water to reduce the propane temperature from 150°F to its condensation temperature of 105.2°F is

$$Q = \dot{m}c_p(T_1 - T_2) = \left(2.7 \times 10^6 \frac{\text{lbm}}{\text{hr}}\right)\left(0.39 \frac{\text{Btu}}{\text{lbm-}°F}\right)$$
$$\times (150°F - 105.2°F)$$
$$= 4.72 \times 10^7 \text{ Btu/hr}$$

The heat transferred during the condensation of propane is

$$Q_{\text{cond}} = \dot{m}\lambda = \left(2.7 \times 10^6 \frac{\text{lbm}}{\text{hr}}\right)\left(138.1 \frac{\text{Btu}}{\text{lbm}}\right)$$
$$= 3.73 \times 10^8 \text{ Btu/hr}$$

The water needs to remove enough heat from the propane to accomplish the condensation. The total heat transferred from the propane to the water is

$$Q_{\text{duty}} = Q + Q_{\text{cond}}$$
$$= 4.72 \times 10^7 \frac{\text{Btu}}{\text{hr}} + 3.73 \times 10^8 \frac{\text{Btu}}{\text{hr}}$$
$$= 4.20 \times 10^8 \text{ Btu/hr}$$

Because the condensation process is to be treated as approximately isothermal at 105.2°F, the logarithmic mean temperature difference is

$$\Delta T_{\text{lm}} = \frac{(T_2 - T_1') - (T_2 - T_2')}{\ln \frac{T_2 - T_1'}{T_2 - T_2'}}$$
$$= \frac{(105.2°F - 86°F) - (105.2°F - 96°F)}{\ln \frac{105.2°F - 86°F}{105.2°F - 96°F}}$$
$$= 13.59°F$$

The heat-transfer area is

$$A_o = \frac{Q_{\text{duty}}}{U_o \Delta T_{\text{lm}}} = \frac{4.20 \times 10^8 \frac{\text{Btu}}{\text{hr}}}{\left(95.9 \frac{\text{Btu}}{\text{ft}^2\text{-hr-}°F}\right)(13.59°F)}$$
$$= 3.22 \times 10^5 \text{ ft}^2 \quad (3.2 \times 10^5 \text{ ft}^2)$$

The answer is (C).

Why Other Options Are Wrong

(A) This incorrect answer uses the heat of condensation of propane instead of the total duty when calculating the area.

(B) This incorrect answer omits the heat of cooling the gaseous propane when calculating the area.

(D) This incorrect answer uses the difference between the inlet and outlet temperatures of the water instead of the logarithmic mean temperature difference when calculating the area.

SOLUTION 53

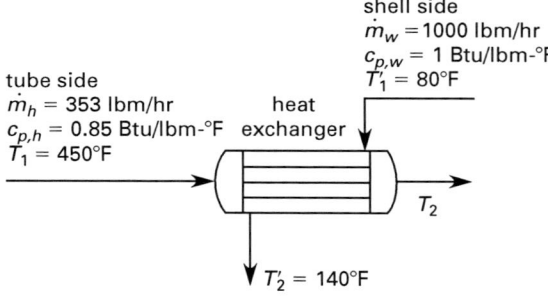

An energy balance around the heat exchanger gives

$$\dot{m}_w c_{p,w}(T'_1 - T'_2) = \dot{m}_h c_{p,h}(T_2 - T_1)$$

$$T_2 = \frac{\dot{m}_w c_{p,w}(T'_1 - T'_2) + \dot{m}_h c_{p,h} T_1}{\dot{m}_h c_{p,h}}$$

$$= \frac{\left(1000 \,\frac{\text{lbm}}{\text{hr}}\right)\left(1 \,\frac{\text{Btu}}{\text{lbm-°F}}\right)(80°\text{F} - 140°\text{F})}{\left(353 \,\frac{\text{lbm}}{\text{hr}}\right)\left(0.85 \,\frac{\text{Btu}}{\text{lbm-°F}}\right)(450°\text{F})}$$

$$+ \frac{\left(353 \,\frac{\text{lbm}}{\text{hr}}\right)\left(0.85 \,\frac{\text{Btu}}{\text{lbm-°F}}\right)(450°\text{F})}{\left(353 \,\frac{\text{lbm}}{\text{hr}}\right)\left(0.85 \,\frac{\text{Btu}}{\text{lbm-°F}}\right)}$$

$$= 250°\text{F}$$

The logarithmic mean temperature difference is

$$\Delta T_{\text{lm}} = \frac{(T_2 - T'_1) - (T_1 - T'_2)}{\ln \dfrac{T_2 - T'_1}{T_1 - T'_2}}$$

$$= \frac{(250°\text{F} - 80°\text{F}) - (450°\text{F} - 140°\text{F})}{\ln \dfrac{250°\text{F} - 80°\text{F}}{450°\text{F} - 140°\text{F}}}$$

$$= 233.03°\text{F} \quad (230°\text{F})$$

The answer is (C).

Why Other Options Are Wrong

(A) When calculating the logarithmic mean temperature difference, this incorrect answer exchanges the temperature of the effluent water with that of the effluent organic fluid.

(B) When calculating the logarithmic mean temperature difference, this incorrect answer exchanges the temperature of the effluent organic fluid with the temperature of the organic fluid entering the tubes.

(D) When calculating the effluent organic fluid temperature, this incorrect answer exchanges the temperature of the water in the shell with the temperature of the organic fluid in the tubes.

SOLUTION 54

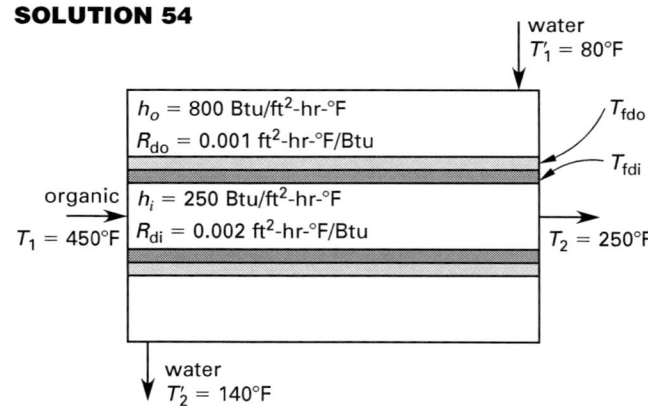

Fluid properties should be evaluated at the bulk temperature. The bulk temperature for the tube side is

$$T_{\text{bulk,tube}} = \frac{T_1 + T_2}{2} = \frac{450°\text{F} + 250°\text{F}}{2}$$
$$= 350°\text{F}$$

The bulk temperature for the shell side is

$$T_{\text{bulk,shell}} = \frac{T'_1 + T'_2}{2} = \frac{80°\text{F} + 140°\text{F}}{2}$$
$$= 110°\text{F}$$

The temperature difference is

$$\Delta T = T_{\text{bulk,tube}} - T_{\text{bulk,shell}} = 350°\text{F} - 110°\text{F}$$
$$= 240°\text{F}$$

The rate of heat transfer between the water, in the shell, and the film formed on the surface of the shell, Q, is

$$Q = U_o A_o \Delta T = h_o (T_{\text{fdo}} - T_{\text{bulk,shell}}) \quad \text{[I]}$$

Let ΔT be defined as

$$\Delta T_o = T_{\text{fdo}} - T_{\text{bulk,shell}}$$

Replacing in Eq. I gives

$$Q = h_o A_o \Delta T_o$$

The outside tube radius is r_o. The inside tube radius is r_i. The inverse of the overall heat-transfer coefficient based on the outside area is

$$\frac{1}{U_o} = \frac{1}{h_o} + \frac{r_o}{h_i r_i} + R_{\text{di}} + R_{\text{do}}$$

Solving for ΔT_o and assuming r_o is approximately equal to r_i,

$$\Delta T_o = \left(\frac{\frac{1}{h_o}}{\frac{r_o}{h_i r_i} + \frac{1}{h_o} + R_{\text{di}} + R_{\text{do}}} \right) \Delta T$$

$$= \left(\frac{\frac{1}{800 \ \frac{\text{Btu}}{\text{ft}^2\text{-hr-}°\text{F}}}}{\frac{1}{250 \ \frac{\text{Btu}}{\text{ft}^2\text{-hr-}°\text{F}}} + \frac{1}{800 \ \frac{\text{Btu}}{\text{ft}^2\text{-hr-}°\text{F}}} + 0.002 \ \frac{\text{ft}^2\text{- hr-}°\text{F}}{\text{Btu}} + 0.001 \ \frac{\text{ft}^2\text{-hr-}°\text{F}}{\text{Btu}}} \right)$$
$$\times (240°\text{F})$$

$$= 36.36°\text{F}$$

The temperature of the fouling in contact with the organic layer is

$$T_{\text{fdo}} = T_{\text{bulk,shell}} + \Delta T_o$$
$$= 110°\text{F} + 36.36°\text{F}$$
$$= 146.36°\text{F}$$

The rate of heat transfer between the organic, in the tubes, and the film formed on the surface of the tubes, Q, is

$$Q = U_i A_i \Delta T$$
$$= h_i A_i (T_{\text{bulk,tube}} - T_{\text{fdi}}) \quad [\text{II}]$$

Let ΔT_i be defined as

$$\Delta T_i = T_{\text{bulk,tube}} - T_{\text{fdi}}$$

Replacing in Eq. II gives

$$Q = h_i A_i \Delta T_i$$

The inverse of the overall heat-transfer coefficient based on the inside area is

$$\frac{1}{U_i} = \frac{1}{h_i} + \frac{r_i}{h_o r_o} + R_{\text{do}} + R_{\text{di}}$$

The temperature difference on the tube side is

$$\Delta T_i = \frac{\left(\frac{1}{h_i}\right) \Delta T}{\frac{1}{h_i} + \frac{r_i}{h_o r_o} + R_{\text{di}} + R_{\text{do}}}$$

$$= \frac{\left(\frac{1}{250 \ \frac{\text{Btu}}{\text{ft}^2\text{-hr-}°\text{F}}} \right)(240°\text{F})}{\frac{1}{250 \ \frac{\text{Btu}}{\text{ft}^2\text{-hr-}°\text{F}}} + \frac{1}{800 \ \frac{\text{Btu}}{\text{ft}^2\text{-hr-}°\text{F}}}}$$
$$+ 0.001 \ \frac{\text{ft}^2\text{-hr-}°\text{F}}{\text{Btu}} + 0.002 \ \frac{\text{ft}^2\text{-hr-}°\text{F}}{\text{Btu}}$$

$$= 116.36°\text{F}$$

$$T_{\text{fdi}} = T_{\text{bulk,tube}} - \Delta T_i$$
$$= 350°\text{F} - 116.36°\text{F}$$
$$= 233.64°\text{F}$$

From the table given, the viscosity of the organic liquid at $T_{\text{bulk,tube}}$ of 350°F is 0.14 cP. The viscosity of the organic liquid at T_{fdi} of 234°F is 0.45 cP. From the table given, the viscosity of the water at $T_{\text{bulk,shell}}$ of 110°F is 0.65 cP. The viscosity of the water at T_{fdo} of 146°F is 0.47 cP. The viscosity correction factor is

$$\left(\frac{\mu_{110°\text{F}}}{\mu_{146°\text{F}}} \right)^{0.14} = \left(\frac{0.65 \text{ cP}}{0.47 \text{ cP}} \right)^{0.14} = 1.05$$

The heat-transfer coefficient on the shell side, corrected for viscosity, is

$$h_{o,\text{corrected}} = h_o (\text{correction factor})$$
$$= \left(800 \ \frac{\text{Btu}}{\text{ft}^2\text{-hr-}°\text{F}} \right) (1.05)$$
$$= 840 \text{ Btu/hr-ft}^2\text{-}°\text{F}$$

The viscosity correction factor for the tube side, based on the organic liquid, is

$$\left(\frac{\mu_{350°\text{F}}}{\mu_{234°\text{F}}} \right)^{0.14} = \left(\frac{0.14 \text{ cP}}{0.45 \text{ cP}} \right)^{0.14} = 0.849$$

The heat-transfer coefficient on the tube side, corrected for viscosity, is

$$h_{i,\text{corrected}} = h_i (\text{correction factor})$$
$$= \left(250 \ \frac{\text{Btu}}{\text{ft}^2\text{-hr-}°\text{F}} \right) (0.849)$$
$$= 212.3 \text{ Btu/ft}^2\text{-hr-}°\text{F}$$

The overall fouled-clean equivalent heat-transfer coefficient, corrected for viscosity, is

$$U_{o,\text{clean corrected}} = \cfrac{1}{\cfrac{r_o}{h_{i,\text{corrected}} r_i} + \cfrac{1}{h_{o,\text{corrected}}}}$$

$$= \cfrac{1}{\cfrac{1}{212.3 \,\dfrac{\text{Btu}}{\text{ft}^2\text{-hr-}°\text{F}}} + \cfrac{1}{840 \,\dfrac{\text{Btu}}{\text{ft}^2\text{-hr-}°\text{F}}}}$$

$$= 169.5 \text{ Btu/ft}^2\text{-hr-}°\text{F}$$
$$(170 \text{ Btu/ft}^2\text{-hr-}°\text{F})$$

The answer is (B).

Why Other Options Are Wrong

(A) This incorrect answer omits the exponent 0.14 when calculating the correction factor, to correct for viscosity.

(C) This answer does not correct for viscosity when calculating the heat-transfer coefficient.

(D) This incorrect answer uses the opposite sign when calculating the overall fouled-clean coefficient.

SOLUTION 55

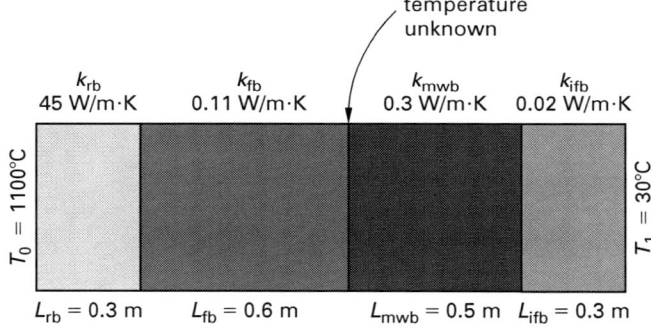

The transversal area of the wall, A, is 1 m^2. The thermal resistance to heat transfer of the refractory brick is

$$R_{\text{rb}} = \frac{L_{\text{rb}}}{k_{\text{rb}} A} = \frac{0.3 \text{ m}}{\left(45 \,\dfrac{\text{W}}{\text{m·K}}\right)(1 \text{ m}^2)}$$

$$= 0.00667 \text{ K/W}$$

The thermal resistance of the firebrick is

$$R_{\text{fb}} = \frac{L_{\text{fb}}}{k_{\text{fb}} A} = \frac{0.6 \text{ m}}{\left(0.11 \,\dfrac{\text{W}}{\text{m·K}}\right)(1 \text{ m}^2)}$$

$$= 5.455 \text{ K/W}$$

The thermal resistance of the mineral wool brick is

$$R_{\text{mwb}} = \frac{L_{\text{mwb}}}{k_{\text{mwb}} A} = \frac{0.5 \text{ m}}{\left(0.3 \,\dfrac{\text{W}}{\text{m·K}}\right)(1 \text{ m}^2)}$$

$$= 1.667 \text{ K/W}$$

The thermal resistance of the insulating firebrick is

$$R_{\text{ifb}} = \frac{L_{\text{ifb}}}{k_{\text{ifb}} A} = \frac{0.3 \text{ m}}{\left(0.02 \,\dfrac{\text{W}}{\text{m·K}}\right)(1 \text{ m}^2)}$$

$$= 15 \text{ K/W}$$

The furnace wall total thermal resistance is

$$R = R_{\text{rb}} + R_{\text{fb}} + R_{\text{mwb}} + R_{\text{ifb}}$$
$$= 0.00667 \,\frac{\text{K}}{\text{W}} + 5.455 \,\frac{\text{K}}{\text{W}} + 1.667 \,\frac{\text{K}}{\text{W}} + 15 \,\frac{\text{K}}{\text{W}}$$
$$= 22.13 \text{ K/W} \quad (22.13°\text{C/W})$$

The heat-flow rate is

$$q = \frac{T_0 - T_1}{R} = \frac{1100°\text{C} - 30°\text{C}}{22.13 \,\dfrac{°\text{C}}{\text{W}}}$$

$$= 48.35 \text{ W}$$

The interface temperature between the firebrick and the mineral wool brick is

$$T = T_0 - q(R_{\text{fb}} + R_{\text{mwb}})$$
$$= 1100°\text{C} + 273°$$
$$\quad - (48.35 \text{ W})\left(5.455 \,\frac{\text{K}}{\text{W}} + 1.667 \,\frac{\text{K}}{\text{W}}\right) - 273°$$
$$= 755.65°\text{C} \quad (800°\text{C})$$

The answer is (C).

Why Other Options Are Wrong

(A) This incorrect answer does not convert the inner wall temperature from degrees Celsius to the absolute scale when calculating the interface temperature. The two terms added are in different temperature scales.

(B) This answer erroneously uses the resistance of the insulating firebrick instead of that of firebrick when calculating the interface temperature.

(D) This incorrect answer is the interface temperature in the absolute scale.

SOLUTION 56

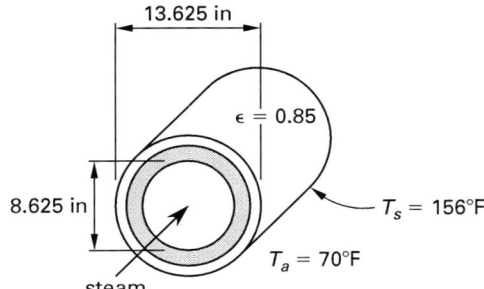

The temperature of the ambient air is
$$T_a = 70°F + 460° = 530°R$$

The surface temperature is
$$T_s = 156°F + 460° = 616°R$$

The external pipe diameter is
$$d_o = 8.625 \text{ in} + 1.5 \text{ in} + 1.5 \text{ in} + 1 \text{ in} + 1 \text{ in}$$
$$= 13.625 \text{ in}$$

For a horizontal pipe, the convection heat-transfer coefficient is given as
$$h_c = \left(0.5 \ \frac{\text{Btu-in}^{0.25}}{\text{ft}^2\text{-hr-}°\text{R}^{1.25}}\right) \left(\frac{T_s - T_a}{d_0}\right)^{0.25}$$
$$= \left(0.5 \ \frac{\text{Btu-in}^{0.25}}{\text{ft}^2\text{-hr-}°\text{R}^{1.25}}\right) \left(\frac{616°R - 530°R}{13.625 \text{ in}}\right)^{0.25}$$
$$= 0.7925 \ \text{Btu/ft}^2\text{-hr-}°\text{R}$$

The Stefan-Boltzmann constant, σ, is 0.173×10^{-8} Btu/ft^2-hr-°R^4. The emissivity of the surface of the pipe, ϵ, is given as 0.85.

The radiation coefficient is
$$h_r = \sigma \left(\frac{\epsilon}{T_s - T_a}\right)(T_s^4 - T_a^4)$$
$$= \left(0.173 \times 10^{-8} \ \frac{\text{Btu}}{\text{ft}^2\text{-hr-}°\text{R}^4}\right)\left(\frac{0.85}{616°R - 530°R}\right)$$
$$\times \left((616°R)^4 - (530°R)^4\right)$$
$$= 1.1128 \ \text{Btu/ft}^2\text{-hr-}°\text{R}$$

The combined coefficient is
$$h_a = h_c + h_r$$
$$= 0.7925 \ \frac{\text{Btu}}{\text{ft}^2\text{-hr-}°\text{R}} + 1.1128 \ \frac{\text{Btu}}{\text{ft}^2\text{-hr-}°\text{R}}$$
$$= 1.9053 \ \text{Btu/ft}^2\text{-hr-}°\text{R} \quad (1.9 \ \text{Btu/ft}^2\text{-hr-}°\text{F})$$

The answer is (B).

Why Other Options Are Wrong

(A) This incorrect answer calculates the radiation heat-transfer coefficient by performing the subtraction of the temperatures in degrees Fahrenheit.

(C) This answer erroneously uses feet instead of inches for the external diameter when calculating the convection heat-transfer coefficient.

(D) This incorrect answer omits the exponent 0.25 when calculating the convection heat-transfer coefficient.

SOLUTION 57

The refrigeration unit cost, C_r, is given as \$36,523. The condenser cost, C_{cond}, is given as \$12,615. The recovery tank cost, C_{tank}, is given as \$3940. The equipment cost is
$$C_{\text{eq}} = C_r + C_{\text{cond}} + C_{\text{tank}}$$
$$= \$36{,}523 + \$12{,}615 + \$3940$$
$$= \$53{,}078$$

The purchased equipment cost is
$$C_{\text{peq}} = 1.18 C_{\text{eq,total}} = (1.18)(\$53{,}078)$$
$$= \$62{,}632$$

The total capital investment is
$$\text{TCI} = 1.74 C_{\text{peq}} = (1.74)(\$62{,}632)$$
$$= \$108{,}980 \quad (\$110{,}000)$$

The answer is (D).

Why Other Options Are Wrong

(A) This incorrect answer omits the purchased equipment cost factor of 1.74 when calculating the total capital investment.

(B) This incorrect answer omits the equipment cost factor of 1.18 when calculating the purchased equipment cost.

(C) This incorrect answer omits the cost of the tank when calculating the equipment cost.

SOLUTION 58

The operating hours per year are
$$t_{\text{op}} = \left(8 \ \frac{\text{hr}}{\text{day}}\right)\left(5 \ \frac{\text{days}}{\text{wk}}\right)\left(52 \ \frac{\text{wk}}{\text{yr}}\right)$$
$$= 2080 \ \text{hr/yr}$$

The annual operating cost is
$$C_{\text{op}} = \left(30 \ \frac{\text{min}}{\text{day}}\right)\left(\frac{1 \ \text{hr}}{60 \ \text{min}}\right)\left(5 \ \frac{\text{days}}{\text{wk}}\right)$$
$$\times \left(52 \ \frac{\text{wk}}{\text{yr}}\right)\left(\frac{\$16.07}{\text{hr}}\right)$$
$$= \$2089/\text{yr}$$

The annual supervisory labor cost is 15% of the operating cost.

$$C_{\text{slb}} = 0.15 C_{\text{op}} = (0.15)\left(\frac{\$2089}{\text{yr}}\right)$$
$$= \$313/\text{yr}$$

The maintenance labor cost per year is

$$\begin{aligned}
C_{\text{mlb}} &= \left(30 \ \frac{\text{min}}{\text{shift}}\right)\left(\frac{1 \ \text{hr}}{60 \ \text{min}}\right)\left(\frac{1 \ \text{shift}}{8 \ \text{hr}}\right) \\
&\quad \times \left(\frac{\$17.50}{\text{hr}}\right)(\text{oph}) \\
&= \left(30 \ \frac{\text{min}}{\text{shift}}\right)\left(\frac{1 \ \text{hr}}{60 \ \text{min}}\right)\left(\frac{1 \ \text{shift}}{8 \ \text{hr}}\right) \\
&\quad \times \left(\frac{\$17.50}{\text{hr}}\right)\left(2080 \ \frac{\text{hr}}{\text{yr}}\right) \\
&= \$2275/\text{yr}
\end{aligned}$$

Because the maintenance labor and the maintenance material costs are equal, the annual maintenance material cost is

$$C_{\text{mmt}} = C_{\text{mlb}} = \$2275/\text{yr}$$

The refrigeration unit capacity, c, is given as 14.1 tons. The power, P, is given as 4.4 kW/ton. The electricity requirement cost, C_{relc}, is given as \$0.0461/kW·hr. The compressor efficiency, η, is given as 0.85. The annual electricity cost is therefore

$$\begin{aligned}
C_{\text{elc}} &= \frac{cPC_{\text{relc}}t_{\text{op}}}{\eta} \\
&= \frac{(14.1 \ \text{tons})\left(4.4 \ \frac{\text{kW}}{\text{ton}}\right)\left(\frac{\$0.0461}{\text{kW·h}}\right)\left(2080 \ \frac{\text{hr}}{\text{yr}}\right)}{0.85} \\
&= \$6999/\text{yr}
\end{aligned}$$

The total direct cost per year is

$$\begin{aligned}
C_{\text{dir,total}} &= C_{\text{op}} + C_{\text{slb}} + C_{\text{mlb}} + C_{\text{mmt}} + C_{\text{elc}} \\
&= \frac{\$2089}{\text{yr}} + \frac{\$313}{\text{yr}} + \frac{\$2275}{\text{yr}} \\
&\quad + \frac{\$2275}{\text{yr}} + \frac{\$6999}{\text{yr}} \\
&= \$13{,}951/\text{yr} \quad (\$14{,}000/\text{yr})
\end{aligned}$$

The answer is (D).

Why Other Options Are Wrong

(A) This incorrect answer omits the cost of the operating labor and the maintenance material when calculating the total annual direct cost.

(B) This incorrect answer omits the cost of the maintenance labor or the maintenance material when calculating the total annual direct cost.

(C) This incorrect answer omits the supervisory cost when calculating the total annual direct cost.

SOLUTION 59

The total capital investment, TCI, is given as \$110,873/yr. Because the capital recovery cost is 0.1098 times the total capital investment, the annual capital recovery cost is

$$\begin{aligned}
C_{\text{cr}} &= \left(\frac{0.1098}{\text{yr}}\right)\text{TCI} = \left(\frac{0.1098}{\text{yr}}\right)(\$110{,}873) \\
&= \$12{,}174/\text{yr}
\end{aligned}$$

The total annual maintenance cost is the sum of the operating labor and maintenance material costs.

$$\begin{aligned}
C_{m,\text{total}} &= C_{\text{olb}} + C_{\text{mmt}} \\
&= \frac{\$2314}{\text{yr}} + \frac{\$2314}{\text{yr}} \\
&= \$4628/\text{yr}
\end{aligned}$$

The annual operating cost, C_{op}, is given as \$2230/yr. The total annual labor and maintenance cost is the sum of the operating cost and the total maintenance cost.

$$C_{\text{lbmn,total}} = C_{m,\text{total}} + C_{\text{op}} = \frac{\$4628}{\text{yr}} + \frac{\$2230}{\text{yr}}$$
$$= \$6858/\text{yr}$$

Because the overhead cost is 61% of the total labor and maintenance cost, the overhead cost per year is

$$C_{\text{over}} = 0.61 C_{\text{lbmn,total}} = (0.61)\left(\frac{\$6858}{\text{yr}}\right)$$
$$= \$4183/\text{yr}$$

The other yearly indirect costs are comprised of 4% of the total capital investment.

$$\begin{aligned}
C_{\text{oi}} &= \left(\frac{0.04}{\text{yr}}\right)\text{TCI} = \left(\frac{0.04}{\text{yr}}\right)(\$110{,}873) \\
&= \$4435/\text{yr}
\end{aligned}$$

The indirect annual cost is the sum of the capital recovery cost, overhead, and other indirect costs.

$$\begin{aligned}
C_{\text{ind}} &= C_{\text{cr}} + C_{\text{over}} + C_{\text{oi}} \\
&= \frac{\$12{,}174}{\text{yr}} + \frac{\$4183}{\text{yr}} + \frac{\$4435}{\text{yr}} \\
&= \$20{,}792/\text{yr} \quad (\$20{,}800/\text{yr})
\end{aligned}$$

The answer is (D).

Why Other Options Are Wrong

(A) This incorrect answer adds only the capital recovery cost and the overhead cost when calculating the total indirect annual cost. This answer leaves out the OIC from the calculation.

(B) This incorrect answer omits the overhead cost when calculating the total indirect annual cost.

(C) This incorrect answer excludes the total labor cost when calculating the total annual direct cost.

SOLUTION 60

The annual operating cost, C_{op}, is given as $2452/yr, and the annual supervisory labor cost is 15% of the operating cost.

$$C_{slb} = 0.15 C_{op} = (0.15)\left(\frac{\$2452}{yr}\right)$$
$$= \$368/yr$$

The annual maintenance labor cost, C_{mlb}, is given as $2545/yr. The annual maintenance material cost, C_{mmt}, is given as $2545/yr. The total annual maintenance cost is the sum of the maintenance labor and maintenance material costs.

$$C_{m,total} = C_{mlb} + C_{mmt} = \frac{\$2545}{yr} + \frac{\$2545}{yr}$$
$$= \$5090/yr$$

The annual electricity cost, C_{elc}, is given as $7078/yr. The total annual direct cost is the sum of the annual operating, supervisory labor, maintenance labor, maintenance material, and electricity costs.

$$C_{dir} = C_{op} + C_{slb} + C_{mlb} + C_{mmt} + C_{elc}$$
$$= \frac{\$2452}{yr} + \frac{\$368}{yr} + \frac{\$2545}{yr}$$
$$+ \frac{\$2545}{yr} + \frac{\$7078}{yr}$$
$$= \$14{,}988/yr$$

The total capital investment, TCI, is given as $103,687. The capital recovery cost is 10.98% of the total capital investment.

$$C_{cr} = \left(\frac{0.1098}{yr}\right) TCI = \left(\frac{0.1098}{yr}\right)(\$103{,}687)$$
$$= \$11{,}385/yr$$

The total annual labor cost, $C_{lb,total}$, is given as $2820/yr. The total annual maintenance cost and total annual labor cost combined is the total labor and maintenance cost.

$$C_{lbmn,total} = C_{m,total} + C_{lb,total} = \frac{\$5090}{yr} + \frac{\$2820}{yr}$$
$$= \$7910/yr$$

Because the overhead cost is 61% of the total labor and maintenance cost, the annual overhead cost is

$$C_{over} = 0.61 C_{lbmn,total} = (0.61)\left(\frac{\$7910}{yr}\right)$$
$$= \$4825/yr$$

Because the other indirect cost is 4% of the total capital investment, the other indirect cost per year is

$$C_{oi} = \left(\frac{0.04}{yr}\right) TCI = \left(\frac{0.04}{yr}\right)(\$103{,}687)$$
$$= \$4147/yr$$

Because the total annual indirect cost is comprised of the capital recovery cost, the overhead, and other indirect costs, the total annual indirect cost is

$$C_{ind} = C_{cr} + C_{over} + C_{oi}$$
$$= \frac{\$11{,}385}{yr} + \frac{\$4825}{yr} + \frac{\$4147}{yr}$$
$$= \$20{,}357/yr$$

The recovered acetone resale value is

$$C_{rec} = \left(600\ \frac{lbm}{hr}\right)\left(8\ \frac{hr}{day}\right)\left(5\ \frac{days}{wk}\right)$$
$$\times \left(52\ \frac{wk}{yr}\right)\left(\frac{\$0.10}{lbm}\right)$$
$$= \$124{,}800/yr$$

The net total annual savings of the recovery operation is

$$TAS = C_{rec} - C_{dir} - C_{ind}$$
$$= \frac{\$124{,}800}{yr} - \frac{\$14{,}988}{yr} - \frac{\$20{,}357}{yr}$$
$$= \$89{,}455/yr \quad (\$89{,}500/yr)$$

The answer is (A).

Why Other Options Are Wrong

(B) This incorrect answer is the recovered acetone value minus the indirect cost. This answer omits the direct cost when calculating the total annual cost.

(C) This incorrect answer is the recovered acetone value minus the direct cost. This answer omits the indirect cost when calculating the total annual cost.

(D) This incorrect answer is the total value of the acetone recovered.

MASS TRANSFER
SOLUTION 61

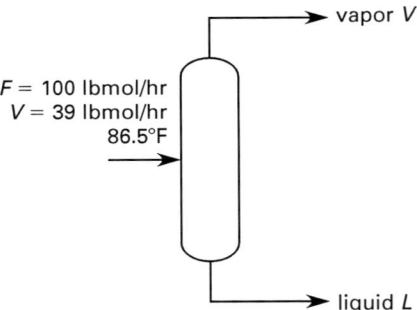

F = 100 lbmol/hr
V = 39 lbmol/hr
86.5°F

vapor V

liquid L

The mole fraction of the vapor stream in the feed is

$$\frac{V}{F} = \frac{39 \, \frac{\text{lbmol}}{\text{hr}}}{100 \, \frac{\text{lbmol}}{\text{hr}}}$$

Applying the partition coefficient definition gives

$$y_i = K_i x_i$$

Rearranging,

$$x_i = \frac{y_i}{K_i} \quad [\text{I}]$$

An overall mass balance around the flash distillation column gives

$$F = L + V$$

Dividing both sides of the preceding equation by F gives

$$1 = \frac{L}{F} + \frac{V}{F}$$

Rearranging,

$$\frac{L}{F} = 1 - \frac{V}{F} \quad [\text{II}]$$

The mass balance of pentane around the flash distillation column is

$$F z_i = L x_i + V y_i \quad [\text{III}]$$

Solving Eq. III for the feed composition,

$$z_i = x_i \left(\frac{L}{F}\right) + y_i \left(\frac{V}{F}\right)$$

Replacing Eq. II into the preceding equation,

$$z_i = x_i \left(1 - \frac{V}{F}\right) + y_i \left(\frac{V}{F}\right)$$

Replacing Eq. I into the preceding equation,

$$z_i = \left(\frac{y_i}{K_i}\right)\left(1 - \frac{V}{F}\right) + y_i \left(\frac{V}{F}\right) \quad [\text{IV}]$$

Solving Eq. IV for the mole fraction of the vapor stream and replacing,

$$y_i = \frac{K_i z_i}{\left(1 - \frac{V}{F}\right) + K_i \left(\frac{V}{F}\right)}$$

$$= \frac{(0.26)(0.340)}{(1 - 0.39) + (0.26)(0.39)}$$

$$= 0.124262$$

The mole fraction of pentane in the liquid phase is

$$x_i = \frac{y_i}{K_i} = \frac{0.124262}{0.26}$$

$$= 0.477931 \quad (0.5)$$

The answer is (C).

Why Other Options Are Wrong

(A) This incorrect answer is the mole fraction of pentane in the vapor phase.

(B) This incorrect answer calculates the mole fraction of propene in the vapor phase instead of calculating it for pentane.

(D) This incorrect answer is the addition of the mole fraction of pentane in the liquid phase and the vapor phase.

SOLUTION 62

The kinematic viscosity of the fluid, ν, is given as 1.384×10^{-5} m^2/s. The density of the fluid, ρ, is given as 1.3 kg/m^3. The porosity of the fluid, ϵ, is given as 0.431. The flow velocity, v, is given as 2.15 m/s. The diameter of the particles, D, is given as 6.35×10^{-3} m. The length of the column, L, is given as 1 m. The modified Reynolds number is

$$\text{Re} = \frac{D\text{v}}{\nu} = \frac{(6.35 \times 10^{-3} \text{ m})\left(2.15 \, \frac{\text{m}}{\text{s}}\right)}{1.384 \times 10^{-5} \, \frac{\text{m}^2}{\text{s}}}$$

$$= 986.45$$

The Ergun equation is

$$\Delta p = \left(\frac{1-\epsilon}{\epsilon^3}\right)\left(\frac{\rho \text{v}^2 L}{D}\right)\left(150 \, \frac{1-\epsilon}{\text{Re}} + 1.75\right)$$

$$= \left(\frac{1 - 0.431}{(0.431)^3}\right)\left(1.3 \, \frac{\text{kg}}{\text{m}^3}\right)\left(\frac{\left(2.15 \, \frac{\text{m}}{\text{s}}\right)^2}{6.35 \times 10^{-3} \text{ m}}\right)$$

$$\times (1 \text{ m}) \left(\frac{(150)(1 - 0.431)}{986} + 1.75\right)$$

$$= 12\,352 \text{ kg/m·s}^2 \quad (12\,000 \text{ Pa})$$

The answer is (D).

Why Other Options Are Wrong

(A) This incorrect answer excludes the pressure drop through the packed column.

(B) This incorrect answer omits the exponent of the porosity in the denominator of the Ergun equation when calculating the pressure drop.

(C) This incorrect answer omits the exponent of the flow velocity in the Ergun equation when calculating the pressure drop.

SOLUTION 63

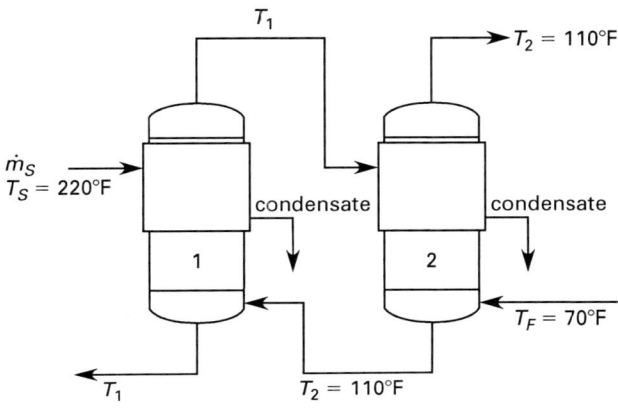

The overall heat-transfer coefficient in the first evaporator, U_1, is given as 450 Btu/ft²-hr-°F. The overall heat-transfer coefficient in the second evaporator, U_2, is given as 315 Btu/ft²-hr-°F. Perform a heat balance. The area of the first evaporator, A_1, is given as 1.8 ft². The area of the second evaporator, A_2, is given as 3.3 ft². The temperature change in the first effect is

$$\Delta T_1 = T_S - T_1$$

The temperature change in the second effect is

$$\Delta T_2 = T_1 - T_2$$

The rate of heat transferred in the first evaporator, q_1, is to be considered approximately equal to the rate of heat transferred in the second evaporator, q_2.

$$q_1 = A_1 U_1 \Delta T_1 = q_2$$
$$= A_2 U_2 \Delta T_2$$
$$A_1 U_1 \Delta T_1 = A_2 U_2 \Delta T_2$$
$$A_1 U_1 (T_S - T_1) = A_2 U_2 (T_1 - T_2)$$

Solving for the temperature in the first effect gives

$$T_1 = \frac{A_1 U_1 T_S + A_2 U_2 T_2}{U_1 A_1 + U_2 A_2}$$

$$= \frac{(1.8 \text{ ft}^2)\left(450 \frac{\text{Btu}}{\text{ft}^2\text{-hr-}°\text{F}}\right)(220°\text{F})}{(1.8 \text{ ft}^2)\left(450 \frac{\text{Btu}}{\text{ft}^2\text{-hr-}°\text{F}}\right)(110°\text{F})}$$
$$+ (3.3 \text{ ft}^2)\left(315 \frac{\text{Btu}}{\text{ft}^2\text{-hr-}°\text{F}}\right)$$

$$= 158.2°\text{F} \quad (160°\text{F})$$

The answer is (B).

Why Other Options Are Wrong

(A) This incorrect answer exchanges the values of the overall heat-transfer coefficients when calculating the temperature in the first effect.

(C) This incorrect answer exchanges the values of the heat-transfer areas when calculating the temperature in the first effect.

(D) When calculating the temperature in the first effect, this incorrect answer exchanges the values of the temperatures of the saturated steam with those of the second effect.

SOLUTION 64

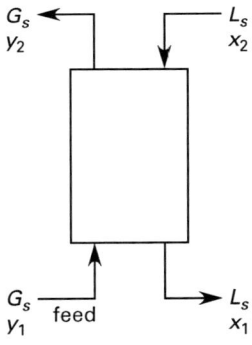

The basis of calculation is 1 lbmol of pollutant-free gas. The number of lbmol of pollutant-free gas, n, is 1. The volume of waste gas, V_g, is given as 10^6 ft³. The volume of HCl entering the absorber, V, is given as 1871 ft³. The number of lbmol of HCl per lbmol of pollutant-free gas is

$$x = n\left(\frac{V}{V_g}\right) = (1 \text{ lbmol})\left(\frac{1871 \text{ ft}^3}{10^6 \text{ ft}^3}\right)$$
$$= 0.001871 \text{ lbmol}$$

The quantity of HCl can be written in terms of lbmol of HCl per lbmol of pollutant-free gas.

$$y_1 = \frac{x}{n-x} = \frac{0.001871}{1-0.001871}$$
$$= 0.0018745 \text{ lbmol HCl/lbmol HCl-free gas}$$

Once the properties of the waste gas stream entering the absorber are known, the properties of the waste gas stream exiting the absorber and the liquid streams entering and exiting the absorber need to be determined.

The liquid does not contain HCl. The concentration of HCl entering the absorber in the liquid, x_2, is 0 lbmol HCl/lbmol HCl-free solvent.

The maximum HCl concentration in the liquid phase in equilibrium with the HCl entering the column in the gas phase, x_1^*, is given as 0.16 lbmol HCl/lbmol HCl-free solvent. The removal efficiency, η, is given as 0.99. The concentration of HCl in the exiting gas stream is

$$y_2 = y_1(1-\eta)$$
$$= \left(0.0018745 \frac{\text{lbmol HCl}}{\text{lbmol solvent}}\right)(1-0.99)$$
$$= 0.000018745 \text{ lbmol HCl/lbmol solvent}$$

The minimum ratio of the molar flow rate of HCl-free solvent to the molar flow rate of HCl-free gas is

$$\left(\frac{L_s}{G_s}\right)_{min} = \frac{y_1 - y_2}{x_1^* - x_2}$$

$$= \frac{0.00187 \frac{\text{lbmol HCl}}{\text{lbmol HCl-free gas}} - 0.0000187 \frac{\text{lbmol HCl}}{\text{lbmol HCl-free gas}}}{0.16 \frac{\text{lbmol HCl}}{\text{lbmol HCl-free solvent}} - 0 \frac{\text{lbmol HCl}}{\text{lbmol HCl-free solvent}}}$$

$$= 0.0116 \text{ lbmol HCl-free solvent/ lbmol HCl-free gas}$$

The actual ratio of the molar flow rate of HCl-free solvent to the molar flow rate of HCl-free gas is

$$\left(\frac{L_s}{G_s}\right)_{actual} = 1.5 \left(\frac{L_s}{G_s}\right)_{min}$$
$$= (1.5)\left(0.0116 \frac{\text{lbmol HCl-free solvent}}{\text{lbmol HCl-free gas}}\right)$$
$$= 0.0174 \text{ lbmol HCl-free solvent/ lbmol HCl-free gas}$$

The final HCl concentration in the liquid phase is

$$x_1 = \frac{y_1 - y_2}{\left(\frac{L_s}{G_s}\right)_{actual}} + x_2$$

$$= \frac{0.00187 \frac{\text{lbmol HCl}}{\text{lbmol solvent}} - 0.0000187 \frac{\text{lbmol HCl}}{\text{lbmol solvent}}}{0.0174 \frac{\text{lbmol HCl-free solvent}}{\text{lbmol HCl}}}$$

$$+ 0 \frac{\text{lbmol HCl}}{\text{lbmol HCl-free solvent}}$$

$$= 0.106 \text{ lbmol HCl/lbmol solvent}$$

The answer is (B).

Why Other Options Are Wrong

(A) This incorrect answer omits the number of moles of HCl per mole of pollutant-free gas when calculating the final HCl concentration in the liquid stream.

(C) This incorrect answer omits the concentration of HCl in the exiting gas stream when calculating the final HCl concentration in the liquid stream.

(D) This incorrect answer uses the minimum liquid-to-gas ratio instead of the actual liquid-to-gas ratio when calculating the final HCl concentration in the liquid stream.

SOLUTION 65

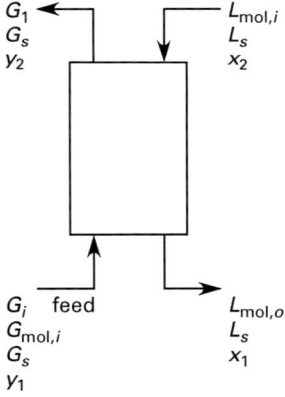

The basis of calculation is the number of moles of pollutant-free gas, n, defined as 1 lbmol. Because the basis of calculation is 1 lbmol of pollutant-free gas, n is 1 lbmol. Because the concentration of HF entering the absorber is 0.18%, the number of lbmol of HF per lbmol of pollutant-free gas, x, is 0.0018 lbmol HF. The concentration of HF entering the absorber in the waste gas can be written in terms of lbmol of HF per lbmol of pollutant-free gas using the following calculation.

$$y_1 = \frac{x}{n-x} = \frac{0.0018}{1-0.0018}$$
$$= 0.001803 \text{ lbmol HF/lbmol HF-free gas}$$

The removal efficiency, η, is given as 0.95. The solvent, a dilute aqueous solution of caustic, is assumed to have the same physical properties as water. Once the properties of the waste gas stream entering the absorber are known, the properties of the waste gas stream exiting the absorber and the liquid streams entering and exiting the absorber must be determined.

Because the solvent does not contain HF, the HF concentration in the liquid entering the absorber, x_2, is given as 0 lbmol HF/lbmol HF-free solvent.

Because the maximum HF concentration in the liquid stream in equilibrium with HF in the gas stream entering the absorber is 0.15 lbmol HF per lbmol of HF-free solvent, the concentration in the liquid phase in equilibrium with the concentration of HF in the gas phase entering the absorber, x_1^*, is 0.15 lbmol HF/lbmol HF-free solvent.

The concentration of HF in the exiting gas stream is

$$y_2 = y_1(1 - \eta)$$
$$= \left(0.001803 \, \frac{\text{lbmol HF}}{\text{lbmol solvent}}\right)(1 - 0.95)$$
$$= 0.0000902 \text{ lbmol HF/lbmol solvent}$$

The density of the gas stream, ρ_G, is given as 0.0709 lbm/ft^3. The minimum ratio of the molar flow rate of HF-free solvent to the molar flow rate of HF-free gas is

$$\left(\frac{L_s}{G_s}\right)_{\text{min}} = \frac{y_1 - y_2}{x_1^* - x_2}$$

$$= \frac{0.001803 \, \frac{\text{lbmol HF}}{\text{lbmol HF-free gas}} - 0.0000902 \, \frac{\text{lbmol HF}}{\text{lbmol HF-free gas}}}{0.15 \, \frac{\text{lbmol HF}}{\text{lbmol HF-free solvent}} - 0 \, \frac{\text{lbmol HF}}{\text{lbmol HF-free solvent}}}$$

$$= 0.01142 \text{ lbmol HF-free solvent/lbmol HF-free gas}$$

Because the actual liquid-to-gas ratio is given as 1.6 times the minimum ratio, the actual liquid-to-gas ratio is

$$\left(\frac{L_s}{G_s}\right)_{\text{actual}} = 1.6 \left(\frac{L_s}{G_s}\right)_{\text{min}}$$
$$= (1.6)\left(0.01142 \, \frac{\text{lbmol HF-free solvent}}{\text{lbmol HF-free gas}}\right)$$
$$= 0.0183 \text{ lbmol HF-free solvent/lbmol HF-free gas}$$

The molecular weight of the waste gas, MW_G, is given as 29 lbm/lbmol. The waste gas flow rate entering the absorber, G_i, is given as 23,000 ft^3/min. The molar flow rate of HF-free gas is

$$G_s = \frac{\rho_G G_i}{(\text{MW}_G)(1 + y_1)}$$
$$= \frac{\left(0.0709 \, \frac{\text{lbm}}{\text{ft}^3}\right)\left(23{,}000 \, \frac{\text{ft}^3}{\text{min}}\right)\left(60 \, \frac{\text{min}}{\text{hr}}\right)}{\left(29 \, \frac{\text{lbm}}{\text{lbmol}}\right)(1 + 0.001803)}$$
$$= 3367.8 \text{ lbmol/hr}$$

The molar flow rate of HF-free solvent is

$$L_s = G_s \left(\frac{L_s}{G_s}\right)_{\text{actual}}$$
$$= \left(3367.8 \, \frac{\text{lbmol}}{\text{hr}}\right)\left(0.0183 \, \frac{\text{lbmol HF-free solvent}}{\text{lbmol HF-free gas}}\right)$$
$$= 61.63 \text{ lbmol/hr}$$

The gas total molar flow rate is

$$G_{\text{mol},i} = G_s(1 + y_1)$$
$$= \left(3367.8 \, \frac{\text{lbmol}}{\text{hr}}\right)(1 + 0.001803)$$
$$= 3373.87 \text{ lbmol/hr}$$

The liquid total molar flow rate is

$$L_{\text{mol},i} = L_s(1 + x_2) = \left(61.63 \, \frac{\text{lbmol}}{\text{hr}}\right)(1 + 0)$$
$$= 61.63 \text{ lbmol/hr}$$

From the equilibrium line, the slope, m, is 0.00104. The absorption factor is

$$\text{AF} = \left(\frac{1}{m}\right)\left(\frac{L_{\text{mol},i}}{G_{\text{mol},i}}\right)$$
$$= \left(\frac{1}{0.00104}\right)\left(\frac{61.63 \, \frac{\text{lbmol}}{\text{hr}}}{3373.87 \, \frac{\text{lbmol}}{\text{hr}}}\right)$$
$$= 17.56 \quad (18)$$

The answer is (C).

Why Other Options Are Wrong

(A) This incorrect answer omits the slope of the operating line when calculating the absorption factor.

(B) This incorrect answer uses the minimum liquid-to-gas ratio instead of the actual value of 1.6 when calculating the absorption factor.

(D) This incorrect answer uses an incorrect expression, $AF = G_{\text{mol},i}/mL_{\text{mol},i}$, to calculate the absorption factor.

SOLUTION 66

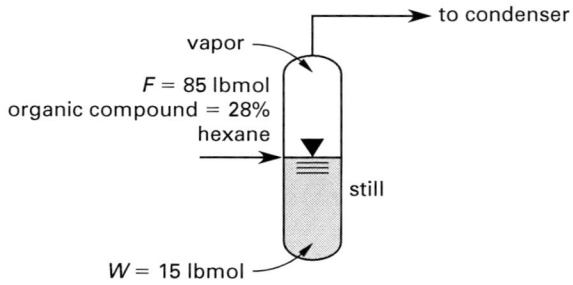

vapor
to condenser
$F = 85$ lbmol
organic compound $= 28\%$
hexane
still
$W = 15$ lbmol

The relative volatility, α, is given as 3.2. Because the feed is 28% organic compound, the initial mole fraction of the organic compound in the liquid phase, x_i, is 0.28. Because this process is a flash distillation, the Rayleigh equation applies. The Rayleigh equation is

$$\ln \frac{F}{W} = \left(\frac{1}{\alpha - 1}\right)\left(\ln \frac{x_i}{x_f} - \alpha \ln \frac{1-x_i}{1-x_f}\right)$$

Replacing,

$$\ln \frac{85 \text{ lbmol}}{15 \text{ lbmol}} = \left(\frac{1}{3.2-1}\right)\left(\ln \frac{0.28}{x_f} - (3.2) \times \left(\ln \frac{1-0.28}{1-x_f}\right)\right)$$

$$= 1.7346 \quad \text{[I]}$$

Solving the preceding equation for x_f using a programmable calculator gives

$$x_f = 0.01671 \quad (2.0 \times 10^{-2})$$

The answer is (D).

Alternate Solution

Equation I could also be resolved iteratively by giving values to x_f. Give values to x_f until the left-side value of Eq. I is 1.7346. For simplification, let Φ be defined as

$$\Phi = \left(\frac{1}{3.2-1}\right)\left(\ln \frac{0.28}{x_f} - (3.2)\left(\ln \frac{1-0.28}{1-x_f}\right)\right)$$

x_f	Φ
0.8	-2.34
0.3	0.072
0.1	0.7926
0.05	1.1863
0.01671	1.7346

Why Other Options Are Wrong

(A) This answer uses the Rayleigh equation incorrectly. This answer applies the following equation.

$$\ln \frac{F}{W} = \left(\frac{1}{\alpha + 1}\right)\left(\ln \frac{x_i}{x_f} - \alpha \ln \frac{1-x_i}{1-x_f}\right)$$

(B) This answer uses the Rayleigh equation incorrectly. This answer applies the following equation.

$$\ln \frac{F}{W} = \left(\frac{1}{\alpha - 1}\right)\left(\ln \frac{x_i}{x_f} + \alpha \ln \frac{1-x_i}{1-x_f}\right)$$

There are two answers. One answer is $x_f = 0.002169$. The second answer is $x_f = 0.8453$.

(C) This answer uses the Rayleigh equation incorrectly. This answer applies the following equation.

$$\ln \frac{F}{W} = \left(\frac{1}{\alpha - 1}\right)\left(\ln \frac{x_i}{x_f} - \alpha \ln \frac{1+x_i}{1-x_f}\right)$$

SOLUTION 67

The total pressure, p, is given as 760 mm Hg. MEK is component 1 and water is component 2.

Because Raoult's law applies, the equilibrium curve is

$$y_i = \frac{x_i p_i^0}{p}$$

The sum of the partial pressure of each component in the vapor phase is

$$p_{\text{total}} = x_F p_1^0 + (1 - x_F) p_2^0$$

For MEK, the Antoine equation is

$$\ln p_1^0 = A - \frac{B}{C + T}$$

$$= 14.2173 - \frac{2831.82\text{K}}{-57.3831\text{K} + T}$$

For water, the Antoine equation is

$$\ln p_2^0 = 18.3036 - \frac{3816.44\text{K}}{-46.13\text{K} + T}$$

Using the two preceding Antoine equations, the calculation of the vapor pressure of each component at different temperatures gives the total pressure. Because the total pressure is 760 mm Hg, the following table shows the vapor pressure of each component, as found by trial and error.

T (K)	p_1^0 (mm Hg)	p_2^0 (mm Hg)	$x_f p_1^0$ (mm Hg)	$(1-x_f)p_2^0$ (mm Hg)	p_{total} (mm Hg)
366	154.7	585.4	61.88	351.3	413.1
376	206.3	840.6	82.53	504.3	586.9
386	270.4	1181	108.1	708.9	817.0
383.76	254.9	1097	102.0	658.0	760.0

The relative volatility is the ratio of the vapor pressure of component 1 to the vapor pressure of component 2. When the total pressure equals 760 mm Hg, the relative volatility is

$$\alpha = \frac{p_1^0}{p_2^0} = \frac{254.9 \text{ mm Hg}}{1097 \text{ mm Hg}}$$
$$= 0.2324 \quad (0.23)$$

The answer is (A).

Why Other Options Are Wrong

(B) This incorrect answer uses 60% acetone instead of 40% when calculating the total pressure.

(C) This incorrect answer does not use Raoult's law when calculating the total pressure.

(D) This incorrect answer divides the pressure of water over the pressure of MEK instead of the pressure of MEK over the pressure of water when calculating α.

SOLUTION 68

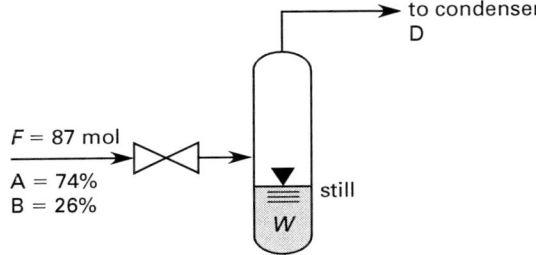

Because the feed contains 74% of solvent A, the mole fraction, x_i, is 0.74. Because the still is distilled with a vaporization of 23%, the liquid content of the still is

$$W = 0.23F = (0.23)(87 \text{ mol})$$
$$= 20.01 \text{ mol}$$

The relative volatility, α, is given as 2.7.

Because the distillation is batch and differential, the Rayleigh equation is

$$\ln\frac{F}{W} = \left(\frac{1}{\alpha - 1}\right)\left(\ln\frac{x_i}{x_f} - \alpha \ln\frac{1 - x_i}{1 - x_f}\right)$$

Solving this equation using a programmable calculator gives

$$x_f = 0.4529$$

The preceding equation could also be resolved iteratively by giving values to x_f. Give values to x_f until the left-side value is 1.470. For simplification, let Φ be defined as

$$\Phi = \left(\frac{1}{2.7 - 1}\right)\left(\ln\frac{0.74}{x_f} - 2.7 \ln\frac{1 - 0.74}{1 - x_f}\right)$$

Solving for x_f,

$\ln\dfrac{87 \text{ mol}}{20.01 \text{ mol}}$	x_f	Φ
1.470	0.8	-0.463
1.470	0.6	0.808
1.470	0.5	1.269
1.470	0.4529	1.470

$$x_f = 0.4529$$

The total number of moles distilled is

$$D = F - W = 87 \text{ mol} - 20.01 \text{ mol}$$
$$= 66.99 \text{ mol}$$

The number of moles of solvent A that remain in the still is

$$S_{A,1} = W x_f = (20.01 \text{ mol})(0.4529)$$
$$= 9.06 \text{ mol}$$

To start the second distillation, the still is charged with solvent B so that the total number of moles in the still is 87 mol. The number of moles of solvent B added to the still at the beginning of the second distillation is 66.99 mol. The liquid that remains in the still after the second distillation is 23% (by mole) of the charge or

$$W = 0.23F = (0.23)(87 \text{ mol})$$
$$= 20.01 \text{ mol}$$

Before the pure solvent B is added to the still to perform the second distillation, the still contains 9.06 mol of A and 10.95 mol of B. The number of moles of B that remains in the still after the first distillation is

$$S_{B,1} = W - S_{A,1} = 20.01 \text{ mol} - 9.06 \text{ mol}$$
$$= 10.95 \text{ mol}$$

Because the feed is 26% solvent B, the number of moles of B in the feed is

$$F_{B,1} = 0.26F = (0.26)(87 \text{ mol})$$
$$= 22.62 \text{ mol}$$

The number of moles of solvent B distilled in the first distillation is

$$D_{B,1} = F_{B,1} - S_{B,1} = 22.62 \text{ mol} - 10.95 \text{ mol}$$
$$= 11.67 \text{ mol}$$

The mole fraction of solvent A in the still at the beginning of the second distillation is

$$x_{i,2} = \frac{S_{A,1}}{F} = \frac{9.06 \text{ mol}}{87 \text{ mol}}$$
$$= 0.10414$$

Applying the Rayleigh equation to the second distillation,

$$\ln \frac{87 \text{ mol}}{20.01 \text{ mol}} = \left(\frac{1}{2.7 - 1}\right) \times \left(\ln \frac{0.10414}{x_{f,2}} - 2.7 \ln \frac{1 - 0.10414}{1 - x_{f,2}}\right)$$

Solving the preceding equation using a programmable calculator gives

$$x_{f,2} = 0.01118$$

The preceding equation could also be resolved iteratively by giving values to x_f. Give values to x_f until the left-side value is 1.470. For simplification, let Ψ be defined as

$$\Psi = \left(\frac{1}{2.7 - 1}\right)\left(\ln \frac{0.10414}{x_{f,2}} - 2.7 \ln \frac{1 - 0.10414}{1 - x_{f,2}}\right)$$

$\ln \frac{87 \text{ mol}}{20.01 \text{ mol}}$	x_f	Ψ
1.470	0.8	−3.581
1.470	0.3	−1.014
1.470	0.01	1.537
1.470	0.01118	1.470

$$x_{f,2} = 0.01118$$

The number of moles of solvent A that remains in the still after the second distillation is

$$S_{A,2} = x_{f,2} W$$
$$= (0.01118)(20.01 \text{ mol})$$
$$= 0.2237 \text{ mol}$$

The number of moles of solvent B that remains in the still after the second distillation is

$$S_{B,2} = W - S_{A,2}$$
$$= 20.01 \text{ mol} - 0.2237 \text{ mol}$$
$$= 19.7863 \text{ mol}$$

The total number of moles of solvent B that co-distills with solvent A (in the second distillation) is

$$D_{B,2} = D - S_{B,2}$$
$$= 66.99 \text{ mol} - 19.7863 \text{ mol}$$
$$= 47.2037 \text{ mol}$$

The total number of moles of solvent B that is collected in the first distillation plus the number of moles of B collected after the second distillation is

$$D_{B,\text{total}} = D_{B,1} + D_{B,2}$$
$$= 11.67 \text{ mol} + 47.2037 \text{ mol}$$
$$= 58.8737 \text{ mol} \quad (60 \text{ mol})$$

The answer is (B).

Why Other Options Are Wrong

(A) This incorrect answer omits the number of moles of solvent removed in the first distillation when calculating the total number of moles of solvent distilled after the second distillation.

(C) This incorrect answer is the number of moles of distillate produced in each of the batches.

(D) This incorrect answer adds the number of moles that remain in the still after the first distillation instead of subtracting when calculating the number of moles of solvent distilled after the second distillation.

SOLUTION 69

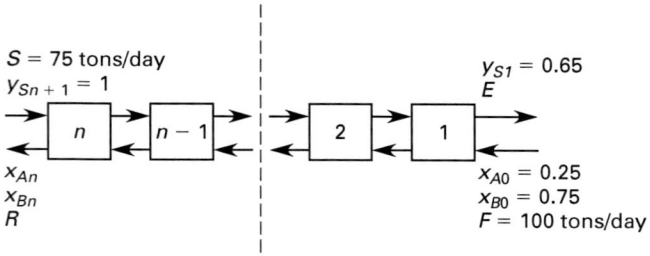

Because the final extract produced contains 35% oil, the mass fraction of oil in the final extract, y_{A1}, is 0.35. Because the final extract produced contains 65% solvent, the mass fraction of naphtha in the final extract, y_{S1}, is 0.65. Because the oil is extracted with pure naphtha, the mass fraction of naphtha entering the system, y_{Sn+1}, is 1.

Because the feed contains 75% sand, the mass flow rate of sand in the feed is

$$F_B = \left(100 \, \frac{\text{tons}}{\text{day}}\right)(0.75) = 75 \text{ tons/day}$$

Because this amount of sand represents 68% of the final raffinate, the mass flow rate of the final raffinate is

$$R = \frac{F_B}{0.68} = \frac{75 \, \frac{\text{tons}}{\text{day}}}{0.68}$$
$$= 110.29 \text{ tons/day}$$

The overall mass balance of all the streams entering and leaving the system gives the mass flow rate of the extract. The feed and the solvent are the streams entering the system, and the raffinate and extract are the streams leaving the system.

$$F + S = R + E$$

Rearranging to solve for E gives

$$E = 100 \, \frac{\text{tons}}{\text{day}} + 75 \, \frac{\text{tons}}{\text{day}} - 110.29 \, \frac{\text{tons}}{\text{day}}$$
$$= 64.71 \text{ tons/day}$$

Because the mass fraction of oil in the final extract is 0.35, the mass flow rate of oil in stage 1 is

$$A_1 = y_{A1}E = (0.35)\left(64.71\ \frac{\text{tons}}{\text{day}}\right)$$
$$= 22.649\ \text{tons/day}$$

Because the feed contains 25% oil, the mass flow rate of oil in the feed is

$$F_{A0} = 0.25F = (0.25)\left(100\ \frac{\text{tons}}{\text{day}}\right)$$
$$= 25\ \text{tons/day}$$

An oil mass balance around the system gives the mass flow rate of oil in the final raffinate. The mass flow rate of oil in the final raffinate is

$$A_n = F_{A0} - A_1 = 25\ \frac{\text{tons}}{\text{day}} - 22.649\ \frac{\text{tons}}{\text{day}}$$
$$= 2.351\ \text{tons/day}$$

The mass fraction of oil in the final raffinate is

$$x_{An} = \frac{A_n}{R} = \frac{2.351\ \frac{\text{tons}}{\text{day}}}{110.29\ \frac{\text{tons}}{\text{day}}}$$
$$= 0.0213$$

The graphical solution follows.

The composition of the overflow is represented on the diagram by a straight line joining the point with coordinates (1.0, 0.0) and the point with coordinates (0.0, 1.0).

Because the underflow from each unit consists of 32% solution and 68% naphtha, the composition of the underflow is represented on the diagram by a straight line parallel to the overflow line just constructed with intercepts (0.32, 0) on the horizontal axis and (0, 0.32) on the vertical axis.

A line is drawn connecting the point on the overflow line (E_1) with coordinates (0.35, 0.65) and the point representing the feed (F) with coordinates (0.25, 0.0). This is line L_1.

The point representing the pure solvent has coordinates (0.0, 1.0). The point representing the final raffinate (R_n) has coordinates (0.0213, 0.299). A line is drawn connecting the point representing the pure solvent with R_n. This is line L_n.

Extend lines L_1 and L_n to their intersection. The intercept of L_1 and L_n determines the difference point (Δ). In this illustration, the coordinates of the difference point are (0.0492, −1.305).

Join the point with coordinates (0.35, 0.65) with the origin of the graph. This line is O_1, which is a "tie

Illustration for Solution 69

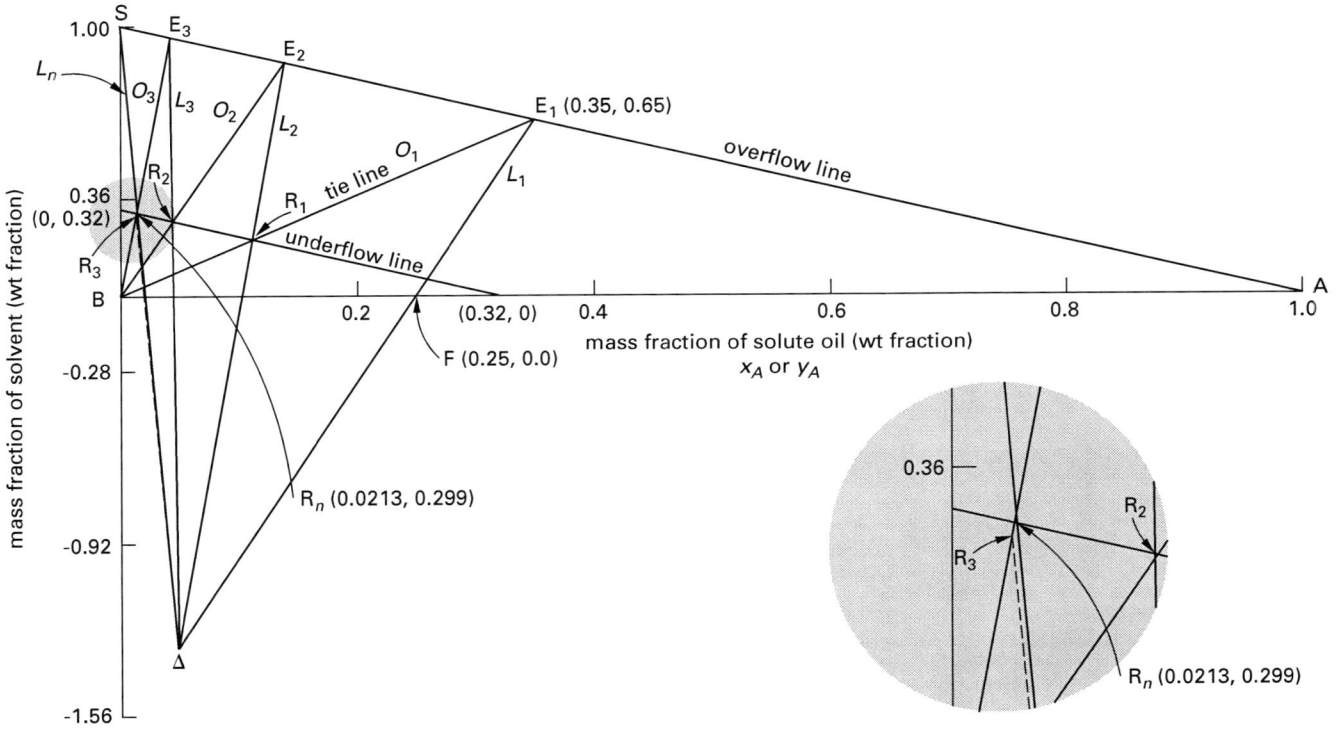

line." This tie line assumes 100% stage efficiency. Line O_1 intercepts the underflow line at point R_1. Join R_1 with Δ. This is line L_2. Extend line L_2 to intercept the overflow line, to determine E_2. From E_2 draw a line to the origin. This line, O_2, is a tie line. This tie line assumes 100% stage efficiency. Line O_2 intercepts the underflow line at R_2. Draw a line from R_2 to Δ. This is line L_3. Extend line L_3 to intercept the overflow line at E_3. Join E_3 with the origin. This line, O_3, is a tie line. This tie line assumes 100% stage efficiency. Line O_3 intercepts the underflow line at R_3. Note that the ordinate of R_3 is to the left of R_n.

The illustration shows the graphical construction to find the number of stages.

It can be seen from the graphical solution that about 2.99 theoretical stages are required for this separation.

The number of stages needed to achieve the desired separation is

$$N = \frac{\text{theoretical number of stages}}{\text{efficiency}} = \frac{2.99}{0.80}$$
$$= 3.74 \quad (4)$$

The answer is (B).

Why Other Options Are Wrong

(A) This incorrect answer excludes the efficiency of the extraction when calculating the number of stages to achieve the desired separation.

(C) This answer erroneously calculates the number of actual stages by adding 1 to the number of theoretical stages and dividing the result by the efficiency.

(D) This answer erroneously calculates the number of actual stages by adding 2 to the number of theoretical stages.

SOLUTION 70

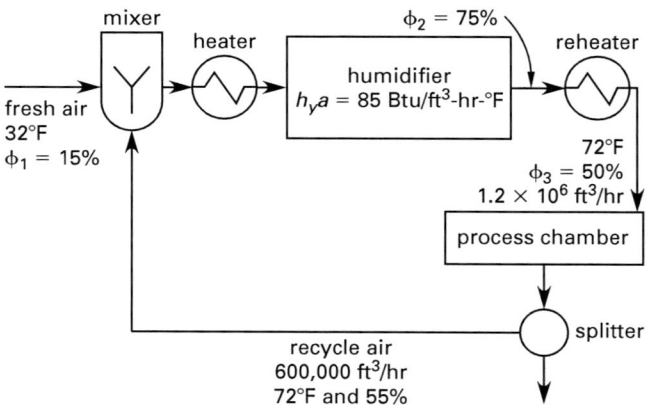

The process chamber air volumetric rate, $\dot{V}_{\text{process}}$, is given as 1.2×10^6 ft^3/hr. The recycle air volumetric rate, $\dot{V}_{\text{recycle}}$, is given as 600,000 ft^3/hr. From the psychrometric chart, for fresh air at 32°F and 15% relative humidity, the specific volume, v_{fresh}, is 12.41 ft^3/lbm dry air. From the psychrometric chart, for recycle air at 72°F and 55% relative humidity, the specific volume, v_{recycle}, is 13.60 ft^3/lbm dry air. From the psychrometric chart, for the process chamber air at 72°F and 50% relative humidity, the specific volume, v_{process}, is 13.58 ft^3/lbm dry air.

The process chamber air mass flow rate is

$$\dot{m}_{\text{process}} = \frac{\dot{V}_{\text{process}}}{v_{\text{process}}} = \frac{1.2 \times 10^6 \, \frac{\text{ft}^3}{\text{hr}}}{13.58 \, \frac{\text{ft}^3}{\text{lbm dry air}}}$$
$$= 88{,}365.2 \text{ lbm dry air/hr}$$

The recycle air mass flow rate is

$$\dot{m}_{\text{recycle}} = \frac{\dot{V}_{\text{recycle}}}{v_{\text{recycle}}} = \frac{600{,}000 \, \frac{\text{ft}^3}{\text{hr}}}{13.60 \, \frac{\text{ft}^3}{\text{lbm dry air}}}$$
$$= 44{,}117.6 \text{ lbm dry air/hr}$$

From a total mass balance around the mixer, the fresh air mass flow rate is

$$\dot{m}_{\text{fresh}} = \dot{m}_{\text{process}} - \dot{m}_{\text{recycle}}$$
$$= 88{,}365.2 \, \frac{\text{lbm dry air}}{\text{hr}} - 44{,}117.6 \, \frac{\text{lbm dry air}}{\text{hr}}$$
$$= 44{,}247.6 \text{ lbm dry air/hr}$$

From the psychrometric chart, the specific humidity of the air at the entrance of the process chamber (72°F, ϕ_3 = 50%), ω_{process}, is 0.0084 lbm water/lbm dry air. Because the splitter does not change the specific humidity of the air, the specific humidity of the recycle air equals the specific humidity of the air in the process chamber.

$$\omega_{\text{recycle}} = \omega_{\text{process}}$$
$$= 0.0084 \text{ lbm water/lbm dry air}$$

The specific humidity of the air at the entrance of the process chamber is equal to the specific humidity of the air at the exit of the humidifier. The heater placed between the adiabatic humidifier and the process chamber will increase the temperature of the air entering without changing its specific humidity. The specific humidity of the air at the exit of the humidifier is

$$\omega_{\text{exit,hum}} = \omega_{\text{process}}$$
$$= 0.0084 \text{ lbm water/lbm dry air}$$

From the psychrometric chart, the temperature of air with a specific humidity of 0.0084 lbm water/lbm dry air and 75% relative humidity, $T_{\text{exit,hum}}$, is 60.5°F. From the psychrometric chart, the saturation temperature of air with a specific humidity of 0.0084 lbm water/lbm dry air and 75% relative humidity, $T_{\text{saturation}}$, is 56°F. From the psychrometric chart, the specific humidity of air at 32°F and 15% relative humidity, ω_{fresh}, is 0.00056 lbm water/lbm dry air. From a water mass balance around the mixer, the specific humidity of the air at the entrance of the humidifier is

$$\omega_{\text{entrance,hum}} = \frac{\dot{m}_{\text{fresh}}\omega_{\text{fresh}} + \dot{m}_{\text{recycle}}\omega_{\text{recycle}}}{\dot{m}_{\text{process}}}$$

$$= \frac{\left(44{,}247.6\ \frac{\text{lbm dry air}}{\text{hr}}\right)\times\left(0.00056\ \frac{\text{lbm water}}{\text{lbm dry air}}\right)}{88{,}365.2\ \frac{\text{lbm dry air}}{\text{hr}}}$$

$$+ \frac{\left(44{,}117.6\ \frac{\text{lbm dry air}}{\text{hr}}\right)\times\left(0.0084\ \frac{\text{lbm water}}{\text{lbm dry air}}\right)}{88{,}365.2\ \frac{\text{lbm dry air}}{\text{hr}}}$$

$$= 0.004474\ \text{lbm water/lbm dry air}$$

From the psychrometric chart, the temperature of air with a specific humidity of 0.0044 lbm water/lbm dry air and a 56°F saturation temperature, $T_{\text{entrance,hum}}$, is 76°F. The following table presents a summary of values.

description	specific volume (ft³/lbm dry air)	temperature (°F)	specific humidity (lbm water/ lbm dry air)	relative humidity (%)
fresh air	12.41	32	0.00056	15
saturation	–	56	–	100
humidifier exit	–	60.5	0.0084	75
recycle	13.60	72	–	55
process chamber	13.58	72	0.0084	50
humidifier entrance	–	76	0.0044	–

The heat capacity of dry air, $c_{p_{\text{air}}}$, is 0.24 Btu/lbm dry air-°F. The heat capacity of the water vapor, $c_{p_{\text{moisture}}}$, is 0.444 Btu/lbm water-°F. The heat capacity of the humid air at the entrance of the humidifier is

$$c_{s,\text{entrance}} = c_{p,\text{air}} + \omega_{\text{entrance,hum}}c_{p,\text{moisture}}$$

$$= 0.24\ \frac{\text{Btu}}{\text{lbm dry air-°F}}$$

$$+ \left(0.004474\ \frac{\text{lbm water}}{\text{lbm dry air}}\right)$$

$$\times \left(0.444\ \frac{\text{Btu}}{\text{lbm water-°F}}\right)$$

$$= 0.2420\ \text{Btu/lbm dry air-°F}$$

The humid heat of the air at the exit of the humidifier is

$$c_{s,\text{exit}} = c_{p,\text{air}} + \omega_{\text{exit,hum}}c_{p,\text{moisture}}$$

$$= 0.24\ \frac{\text{Btu}}{\text{lbm dry air-°F}}$$

$$+ \left(0.0084\ \frac{\text{lbm water}}{\text{lbm dry air}}\right)$$

$$\times \left(0.444\ \frac{\text{Btu}}{\text{lbm water-°F}}\right)$$

$$= 0.2437\ \text{Btu/lbm dry air-°F}$$

The average heat capacity of the air in the humidifier is

$$c_s = \frac{c_{s,\text{entrance}} + c_{s,\text{exit}}}{2}$$

$$= \frac{0.2420\ \frac{\text{Btu}}{\text{lbm dry air-°F}} + 0.2437\ \frac{\text{Btu}}{\text{lbm dry air-°F}}}{2}$$

$$= 0.2429\ \text{Btu/lbm dry air-°F}$$

The heat-transfer coefficient between the air and the surface of the water multiplied by the mass transfer area, $h_y a$, is given as 85 Btu/ft³-hr-°F. Under these conditions the volume required for the humidifier is

$$V = \frac{\left(\ln\dfrac{T_{\text{entrance,hum}} - T_{\text{saturation}}}{T_{\text{exit,hum}} - T_{\text{saturation}}}\right)c_s \dot{m}_{\text{process}}}{h_y a}$$

$$= \frac{\left(\ln\dfrac{76°F - 56°F}{60.5°F - 56°F}\right)\left(0.2429\ \dfrac{\text{Btu}}{\text{lbm dry air·°F}}\right)}{85\ \dfrac{\text{Btu}}{\text{ft}^3\text{-hr-°F}}}$$

$$\times \left(88{,}365.2\ \dfrac{\text{lbm dry air}}{\text{hr}}\right)$$

$$= 376.67\ \text{ft}^3\quad (377\ \text{ft}^3)$$

The answer is (C).

Why Other Options Are Wrong

(A) This incorrect answer uses the mass flow rate of the recycle air instead of the mass flow rate of the air to the process chamber.

(B) This incorrect answer uses the heat capacity of pure dry air instead of the average heat capacity of the mixture used in the process chamber.

(D) This incorrect answer neglects to take the natural logarithm of the term involving the temperatures when calculating the volume.

SOLUTION 71

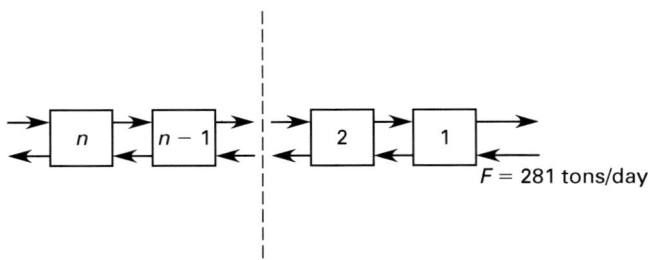

Because the feed contains 10.3% copper sulfate, the mass flow rate of copper sulfate in the feed is

$$A_o = 0.103F = (0.103)\left(281 \; \frac{\text{tons}}{\text{day}}\right)$$
$$= 28.94 \; \text{tons/day}$$

Because the feed contains 85.4% inert, the mass flow rate of inert in the feed is

$$B_o = 0.854F = (0.854)\left(281 \; \frac{\text{tons}}{\text{day}}\right)$$
$$= 239.97 \; \text{tons/day}$$

Because the feed contains 4.3% water, the mass flow rate of water in the feed is

$$S_o = 0.043F = (0.043)\left(281 \; \frac{\text{tons}}{\text{day}}\right)$$
$$= 12.08 \; \text{tons/day}$$

Because the desired recovery of copper sulfate is 92%, the mass flow rate of copper sulfate in the final extract is

$$A_1 = 0.92 A_o = (0.92)\left(28.94 \; \frac{\text{tons}}{\text{day}}\right)$$
$$= 26.62 \; \text{tons/day}$$

Numbering the stages in the process from 1 to n, the mass flow rate of the extract leaving the first step is the mass flow rate of the final extract. Because the final extract produced contains 10% copper sulfate, this value (A_1) represents 10% of the final extract. The extract produced in the first stage is

$$E_1 = \frac{A_1}{0.10} = \frac{26.62 \; \frac{\text{tons}}{\text{day}}}{0.10}$$
$$= 266.2 \; \text{tons/day}$$

The percentage of copper sulfate in the final extract is given as 10%. Because the final extract contains only copper sulfate and water, the percentage of water in the final extract is 90%. The mass flow rate of water in the extract produced in the first stage is

$$S_1 = 0.90 E_1 = (0.90)\left(266.2 \; \frac{\text{tons}}{\text{day}}\right)$$
$$= 239.58 \; \text{tons/day}$$

A mass balance of copper sulfate around the system gives the mass of copper sulfate in the raffinate. The mass of copper sulfate in the feed minus the copper sulfate in the final extract gives the mass flow rate of copper sulfate in the raffinate. The mass flow rate of copper sulfate in the final raffinate is

$$A_n = A_o - A_1 = 28.94 \; \frac{\text{tons}}{\text{day}} - 26.62 \; \frac{\text{tons}}{\text{day}}$$
$$= 2.32 \; \text{tons/day}$$

A mass balance of the inert around the entire system gives the mass flow rate of the inert in the final raffinate. Because the inert stays in the raffinate and the mass flow rate of the inert is constant, the mass flow rate of the inert in the feed equals the mass flow rate of the inert in the final raffinate. Because the inert does not dissolve or change during this process, the mass flow rate of inert in the final raffinate is

$$B_n = B_o = 239.97 \; \text{tons/day}$$

Because the underflow from each unit consists of 66.7% solution and 33.3% inert, the mass flow rate of the solution in the underflow is

$$U = \left(\frac{0.667}{0.333}\right) B_o = \left(\frac{0.667}{0.333}\right)\left(239.97 \; \frac{\text{tons}}{\text{day}}\right)$$
$$= 480.66 \; \text{tons/day}$$

The mass fraction of copper sulfate in the extract solution in stage n equals the mass fraction of copper sulfate in the underflow solution in stage n because the extractor has reached equilibrium conditions in each stage. Because each stage reaches equilibrium conditions, the mass fraction of copper sulfate in any stage in the extract equals the mass fraction of copper sulfate in the raffinate. The solution mass fraction of copper sulfate in stage n is

$$y_{An} = x_{An}$$
$$= \frac{A_o - 0.92 A_o}{U}$$
$$= \frac{28.94 \; \frac{\text{tons}}{\text{day}} - (0.92)\left(28.94 \; \frac{\text{tons}}{\text{day}}\right)}{480.66 \; \frac{\text{tons}}{\text{day}}}$$
$$= 0.004817 \; \text{tons copper sulfate/ton solution}$$

The final solution is composed of water and copper sulfate. This solution is the final extract. A water mass balance around the system gives the mass flow rate of water in the final solution. The mass flow rate of water in the final solution equals the mass flow rate of water in the final extract plus the mass flow rate of water in the final solution minus the mass flow rate of water in the feed. The mass flow rate of water in the final raffinate is

$$W_n = U - A_n = 480.66\ \frac{\text{tons}}{\text{day}} - 2.32\ \frac{\text{tons}}{\text{day}}$$
$$= 478.34\ \text{tons/day}$$

The overall mass balance gives the mass flow rate of fresh water entering the system. The mass flow rate of water entering the system equals the mass flow rate of water in the final extract plus the mass flow rate of water in the final solution minus the water in the feed. A mass balance of water around the system gives

$$W = S_1 + W_n - S_o$$
$$= 239.58\ \frac{\text{tons}}{\text{day}} + 478.34\ \frac{\text{tons}}{\text{day}} - 12.08\ \frac{\text{tons}}{\text{day}}$$
$$= 705.84\ \text{tons/day}$$

The process of determining the number of stages involves the use of a stepping-off technique. Except for the extract solution produced in the first stage (the final extract), the mass flow rate of the extract solution in step $n-1$ equals the mass flow rate of the extract solution in step n. The following relationship is valid for stages 2 to n.

$$E_{n-1} = E_n$$

The extract solution produced in the second stage is

$$E_2 = W = 705.84\ \text{tons/day}$$

Except for the solution produced in the first stage, the following relationship is valid for stages 2 to n.

$$U_{n-1} = U_n$$

The solution mass fraction of copper sulfate in the final product is given as

$$y_1 = x_1 = 0.1$$

A copper sulfate mass balance around stage 1 gives

$$y_1 E_1 + x_1 U_1 = A_o + y_2 E_2$$

Rearranging to solve for y_2 gives

$$y_2 = \frac{y_1 E_1 + x_1 U_1 - A_o}{E_2}$$

$$= \frac{(0.1)\left(266.2\ \frac{\text{tons}}{\text{day}}\right) + (0.1) \times \left(480.66\ \frac{\text{tons}}{\text{day}}\right) - 28.94\ \frac{\text{tons}}{\text{day}}}{705.84\ \frac{\text{tons}}{\text{day}}}$$

$$= 0.06481$$

From stages 2 to n, another set of equations applies.

$$E_i y_i + U_i x_i = E_{i+1} y_{i+1} + U_{i-1} x_{i-1}$$

Because all Us are equal and all Es are equal, solving for y_i gives

$$y_i = \left(\frac{-U}{E}\right) x_i + y_{i+1} + \left(\frac{U}{E}\right) x_{i-1}$$

For stage-wise calculation, replacing $x_i = y_i$ and $x_{i+1} = y_{i+1}$ in the preceding equation gives

$$y_{i+1} = y_i \left(1 + \frac{U}{E}\right) - \left(\frac{U}{E}\right) y_{i-1}$$

The ratio of underflow solution to extract is

$$\frac{U}{E} = \frac{480.66\ \frac{\text{tons}}{\text{day}}}{705.84\ \frac{\text{tons}}{\text{day}}} = 0.68098$$

Replacing this value in the preceding equation gives

$$y_{i+1} = (1 + 0.68098)\, y_i - 0.68098 y_{i-1}$$
$$= 1.68098 y_i - 0.68098 y_{i-1}$$

Starting with the values of y_1 and y_2 previously calculated, use the preceding equation to calculate y_3.

$$y_3 = 1.68098 y_2 - 0.68098 y_1$$
$$= (1.68098)(0.06481) - (0.68098)(0.1)$$
$$= 0.0408$$

With the values of y_2 and y_3, calculate y_4.

$$y_4 = 1.68098 y_3 - 0.68098 y_2$$
$$= (1.68098)(0.0408) - (0.68098)(0.06481)$$
$$= 0.02445$$

With the values of y_3 and y_4, calculate y_5.

$$y_5 = 1.68098 y_4 - 0.68098 y_3$$
$$= (1.68098)(0.02445) - (0.68098)(0.0408)$$
$$= 0.0133$$

With the values of y_4 and y_5, calculate y_6.

$$y_6 = 1.68098 y_5 - 0.68098 y_4$$
$$= (1.68098)(0.0133) - (0.68098)(0.02445)$$
$$= 0.00571$$

With the values of y_5 and y_6, calculate y_7.

$$y_7 = 1.69098 y_6 - 0.68098 y_5$$
$$= (1.69098)(0.00571) - (0.68098)(0.0133)$$
$$= 0.000541$$

Because y_7 gives a value that is smaller than the limit value, x_{An}, the number of stages needed is seven. The following table summarizes the results.

i	y_{i+1}
1	0.06481
2	0.0408
3	0.02444
4	0.0133
5	0.00571
6	0.000541

Seven (7) stages are needed.

The answer is (D).

Why Other Options Are Wrong

(A) This incorrect answer stopped the extraction process too soon. At the end of the third stage, the mass fraction extracted is 0.0408—about 10 times larger than that required in the problem statement of 0.004810.

(B) This incorrect answer stopped the extraction process too soon. At the end of the fifth stage, the mass fraction extracted is 0.0133—larger than that required in the problem statement, which is 0.004810.

(C) This incorrect answer calculates one stage fewer than is needed. At the end of the sixth stage, the mass fraction extracted is 0.00571—larger than that required in the problem statement, which is 0.004810.

SOLUTION 72

Treat the process as if it occurred in two steps. In the first step, the products are formed at a constant temperature of 25°C. In the second step, the temperature of the products increases to 845°C.

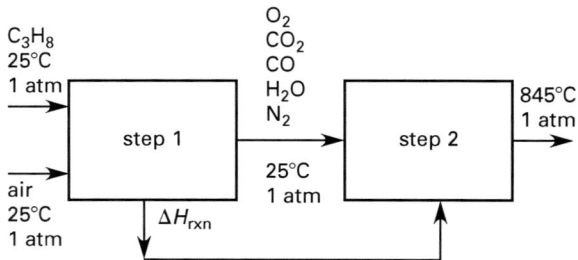

The basis of calculation is 1 mol of C_3H_8 feed. Consider the following reactions to calculate the amount of oxygen and nitrogen in the air supplied. In these reactions, all reactants and products are in the gaseous phase. The balanced reaction for a complete combustion is

$$C_3H_8 + 5O_2 \rightarrow 3CO_2 + 4H_2O$$

The number of moles of oxygen used in the combustion is y. 100 mol of air contain 79 mol of nitrogen and 21 mol of oxygen. Because nitrogen does not react in the combustion, the number of moles of nitrogen in the reactants equals the number of moles of nitrogen in the product. The number of moles of nitrogen on either side of the reaction is

$$n_{N_2} = y\left(\frac{79 \text{ mol}}{21 \text{ mol}}\right)$$

The number of moles of CO_2 produced in the combustion, n_{CO_2} is given as 0.76. The balanced combustion reaction when 76% of C_3H_8 burns to carbon dioxide is

$$C_3H_8 + yO_2 + y\left(\frac{79}{21}\right)N_2$$
$$\rightarrow 0.76 \text{ } CO_2 + (3 - 0.76)\text{ } CO$$
$$+ 4H_2O + y\left(\frac{79}{21}\right)N_2$$

From the balanced combustion reaction, the number of moles of oxygen in the product is

$$n_{O_2,\text{prod}} = y - 0.76 - \frac{3 - 0.76}{2} - \frac{4}{2}$$
$$= y - 3.88$$

The number of moles of nitrogen in the air supplied is

$$n_{N_2,\text{air}} = y\left(\frac{79}{21}\right)$$
$$= 3.76y$$

The stoichiometric table is

component	reactants, n (mol)	products, n (mol)
$C_3H_8(g)$	1	0
$O_2(g)$	y	$y - 3.88$
$CO_2(g)$	0	0.76
$CO(g)$	0	2.24
$H_2O(g)$	0	4
$N_2(g)$	$3.76y$	$3.76y$

Using the stoichiometric coefficients and the enthalpy of formation given, the heat of reaction used to heat products and excess reactants (assuming adiabatic operation) is

$$Q = 0 = \Delta H$$
$$= \Delta H_p + \Delta H_{\text{rxn}}$$

$$\Delta H_{\text{rxn}} = \sum n \Delta H_f$$
$$= n_{\text{products}} \Delta H_{f,\text{products}}$$
$$\quad - n_{\text{reactants}} \Delta H_{f,\text{reactants}}$$
$$= n_{\text{CO}_2} \Delta H_{f_{\text{CO}_2}} + n_{\text{CO}} \Delta H_{f_{\text{CO}}}$$
$$\quad + n_{\text{H}_2\text{O}} \Delta H_{f_{\text{H}_2\text{O}}} - n_{\text{C}_3\text{H}_8} \Delta H_{f_{\text{C}_3\text{H}_8}}$$
$$= (0.76 \text{ mol})\left(-94\,051.8 \, \frac{\text{cal}}{\text{mol}}\right)$$
$$\quad + (2.24 \text{ mol})\left(-26\,415.7 \, \frac{\text{cal}}{\text{mol}}\right)$$
$$\quad + (4 \text{ mol})\left(-57\,798 \, \frac{\text{cal}}{\text{mol}}\right)$$
$$\quad - (1 \text{ mol})\left(-26\,600 \, \frac{\text{cal}}{\text{mol}}\right)$$
$$= -335\,242.5 \text{ cal}$$

The initial temperature is given as

$$T_1 = 25°\text{C} + 273° = 298\text{K}$$

The final temperature is given as

$$T_2 = 845°\text{C} + 273° = 1118\text{K}$$

To calculate the enthalpy change of the products, calculate the enthalpy change of each of the products. The enthalpy change of the oxygen is

$$\Delta H_{p_{\text{O}_2}} = \overline{C}_{p_{\text{O}_2}}(T_2 - T_1)$$
$$= \left(7.83 \, \frac{\text{cal}}{\text{mol} \cdot \text{K}}\right)(1118\text{K} - 298\text{K})$$
$$= 6420.60 \text{ cal/mol}$$

The enthalpy change of the carbon dioxide is

$$\Delta H_{p_{\text{CO}_2}} = \overline{C}_{p_{\text{CO}_2}}(T_2 - T_1)$$
$$= (11.65)(1118\text{K} - 298\text{K})$$
$$= 9553.00 \text{ cal/mol}$$

The enthalpy change of the carbon monoxide is

$$\Delta H_{p_{\text{CO}}} = \overline{C}_{p_{\text{CO}}}(T_2 - T_1)$$
$$= \left(7.49 \, \frac{\text{cal}}{\text{mol} \cdot \text{K}}\right)(1118\text{K} - 298\text{K})$$
$$= 6141.80 \text{ cal/mol}$$

The enthalpy change of the water is

$$\Delta H_{p_{\text{H}_2\text{O}}} = \overline{C}_{p_{\text{H}_2\text{O}}}(T_2 - T_1)$$
$$= \left(9.05 \, \frac{\text{cal}}{\text{mol} \cdot \text{K}}\right)(1118\text{K} - 298\text{K})$$
$$= 7421.00 \text{ cal/mol}$$

The enthalpy change of the nitrogen is

$$\Delta H_{p_{\text{N}_2}} = \overline{C}_{p_{\text{N}_2}}(T_2 - T_1)$$
$$= \left(7.40 \, \frac{\text{cal}}{\text{mol} \cdot \text{K}}\right)(1118\text{K} - 298\text{K})$$
$$= 6068.00 \text{ cal/mol}$$

An energy balance using the enthalpy changes gives the specific heat of the products.

$$\Delta H_p = n_{\text{O}_2} \Delta H_{p_{\text{O}_2}} + n_{\text{CO}_2} \Delta H_{p_{\text{CO}_2}} + n_{\text{CO}} \Delta H_{p_{\text{CO}}}$$
$$\quad + n_{\text{H}_2\text{O}} \Delta H_{p_{\text{H}_2\text{O}}} + n_{\text{N}_2} \Delta H_{p_{\text{N}_2}}$$
$$= (y - 3.88 \text{ mol})\left(6420.60 \, \frac{\text{cal}}{\text{mol}}\right)$$
$$\quad + (0.76 \text{ mol})\left(9553.00 \, \frac{\text{cal}}{\text{mol}}\right)$$
$$\quad + (2.24 \text{ mol})\left(6141.80 \, \frac{\text{cal}}{\text{mol}}\right)$$
$$\quad + (4 \text{ mol})\left(7421.00 \, \frac{\text{cal}}{\text{mol}}\right)$$
$$\quad + (3.76 \text{ mol}) y \left(6068.00 \, \frac{\text{cal}}{\text{mol}}\right)$$
$$= 25\,789.98 \text{ cal} + \left(29\,236.28 \, \frac{\text{cal}}{\text{mol}}\right) y$$

Because the combustor is adiabatic, the specific heat of the products equals the negative of the heat of reaction used to heat the products.

$$\Delta H_p = -\Delta H_{\text{rxn}}$$
$$= 25\,789.98 \text{ cal} + \left(29\,236.28 \, \frac{\text{cal}}{\text{mol}}\right) y$$
$$= 335\,242.54 \text{ cal}$$

Solving for y gives

$$y = \frac{335\,242.54 \text{ cal} - 25\,789.98 \text{ cal}}{29\,236.28 \, \frac{\text{cal}}{\text{mol}}}$$
$$= 10.58 \text{ mol}$$

From the balanced combustion reaction, the theoretical number of moles of oxygen required to complete the reaction is 5 mol. The excess percentage of air is

$$\text{excess \% air} = \frac{\text{actual} - \text{theoretical}}{\text{theoretical}} \times 100\%$$
$$= \frac{10.58 \text{ mol} - 5 \text{ mol}}{5 \text{ mol}} \times 100\%$$
$$= 111.60\% \quad (112\%)$$

The answer is (A).

Why Other Options Are Wrong

(B) This incorrect answer excludes the heat produced by the carbon dioxide, carbon monoxide, and water when calculating the heat produced in the reaction.

(C) This incorrect answer omits the contribution of the water in the products when calculating the heat produced in the reaction.

(D) This answer uses the wrong expression to calculate the theoretical air. This answer omits the subtraction of the theoretical amount of air in the numerator when calculating the excess percentage air.

SOLUTION 73

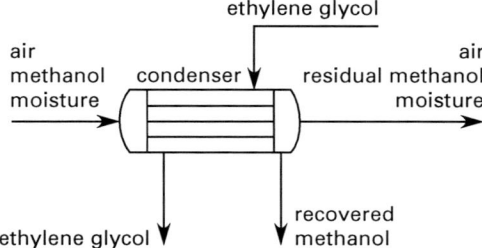

The removal efficiency, η, is given as 0.90. The methanol inlet volume fraction, y, is given as 0.385. Calculate the mass flow rate of methanol recovered and the methanol mass flow rate in the inlet stream. The system temperature is

$$T = 77°F + 460° = 537°R$$

The system pressure, p, is 14.7 lbf/in². The universal gas constant, R, is 10.73 lbf-ft³/lbmol-in²-°R. The heat capacity of the gaseous stream, c_p, is given as 6.95 Btu/lbmol-°F. The inlet stream temperature, T_{in}, is given as 85°C. The condensation temperature of methanol, T_{cond}, is given as 17.3°F. The specific volume of the air/methanol mixture is

$$v = \frac{RT}{p} = \frac{\left(10.73 \ \frac{\text{lbf-ft}^3}{\text{lbmol-in}^2\text{-°R}}\right)(537°R)}{14.7 \ \frac{\text{lbf}}{\text{in}^2}}$$

$$= 392 \ \text{ft}^3/\text{lbmol}$$

The inlet stream flow rate, $\dot{V}$, is given as 200 ft³/min.

The molar flow rate of methanol is

$$F = \left(\frac{\dot{V}}{v}\right)y = \left(\frac{200 \ \frac{\text{ft}^3}{\text{min}}}{392 \ \frac{\text{ft}^3}{\text{lbmol}}}\right)(0.385)\left(60 \ \frac{\text{min}}{\text{hr}}\right)$$

$$= 11.79 \ \text{lbmol/hr}$$

The residual methanol molar flow rate in the outlet stream is

$$F_{out} = F(1-\eta)$$

$$= \left(11.79 \ \frac{\text{lbmol}}{\text{hr}}\right)(1-0.90)$$

$$= 1.179 \ \text{lbmol/hr}$$

The recovered condensed methanol molar flow rate is

$$F_{cond} = F - F_{out}$$

$$= 11.79 \ \frac{\text{lbmol}}{\text{hr}} - 1.179 \ \frac{\text{lbmol}}{\text{hr}}$$

$$= 10.6 \ \text{lbmol/hr}$$

The latent enthalpy for methanol, λ_{T_2}, is given as 17,230.6 Btu/lbmol.

The ethylene glycol needs to remove enough heat to cool the air/methanol mixture from 85°F to its condensation temperature of 17.3°F, and it needs to remove enough heat to condense the methanol. This condensation process is isothermal at 17.3°F.

The molar flow rate of the air/methanol mixture fed to the system is

$$F_{mixture} = \frac{\dot{V}}{v} = \left(\frac{200 \ \frac{\text{ft}^3}{\text{min}}}{392 \ \frac{\text{ft}^3}{\text{lbmol}}}\right)\left(60 \ \frac{\text{min}}{\text{hr}}\right)$$

$$= 30.61 \ \text{lbmol/hr}$$

The heat removed by the ethylene glycol to cool the air/methanol mixture from 85°F to 17.3°F is

$$\Delta H = F_{mixture}c_p(T_{in} - T_{cond})$$

$$= \left(30.61 \ \frac{\text{lbmol}}{\text{hr}}\right)\left(6.95 \ \frac{\text{Btu}}{\text{lbmol-°F}}\right)$$

$$\times (85°F - 17.3°F)$$

$$= 14{,}402.46 \ \text{Btu/hr}$$

During the condensation of methanol, at 17.3°F, the ethylene glycol removes heat.

$$\Delta H_{cond} = F_{cond}\lambda_{T_2}$$

$$= \left(10.6 \ \frac{\text{lbmol}}{\text{hr}}\right)\left(17{,}230.6 \ \frac{\text{Btu}}{\text{lbmol}}\right)$$

$$= 182{,}644.36 \ \text{Btu/hr}$$

The total heat removed by the ethylene glycol is

$$\Delta H_{total} = \Delta H + \Delta H_{cond}$$

$$= 14{,}402.46 \ \frac{\text{Btu}}{\text{hr}} + 182{,}644.36 \ \frac{\text{Btu}}{\text{hr}}$$

$$= 197{,}046.82 \ \text{Btu/hr} \quad (197{,}000 \ \text{Btu/hr})$$

The answer is (D).

Why Other Options Are Wrong

(A) This incorrect answer is the heat removed by the ethylene glycol to bring the air/methanol mixture from 85°F to 17.3°F only. This answer omits the condensation enthalpy change.

(B) This incorrect answer uses as the latent heat the condensation enthalpy at 134°F instead of the enthalpy change at the condensation temperature of 17.3°F when calculating the condensation enthalpy change.

(C) This incorrect answer is the enthalpy change during condensation of methanol. This answer excludes the heat removed by the ethylene glycol during cooling of the air/methanol mixture from 85°F to 17.3°F.

SOLUTION 74

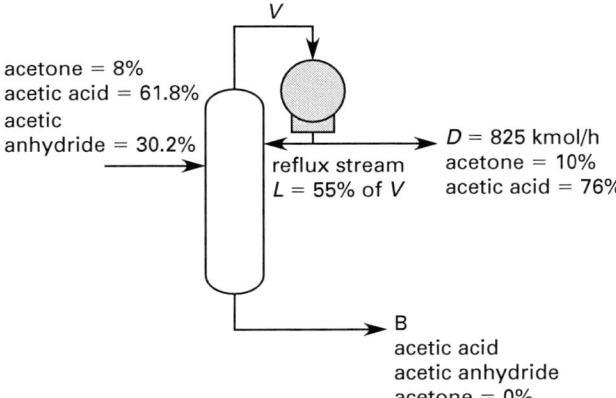

Because the feed is 8 mol% acetone, the mole fraction of acetone in the feed, $x_{\text{ac},F}$, is 0.08. The molar flow rate of the distillate, D, is given as 825 kmol/h. Because the distillate stream contains 10% acetone, the mole fraction of acetone in the distillate stream, $x_{\text{ac},D}$, is 0.10. The mole fraction of acetone in the distillate stream is given as

$$x_{\text{ac},V} = x_{\text{ac},D} = 0.10$$

The mole fraction of acetone in the bottom stream, $x_{\text{ac},B}$, is given as 0. The acetone mole balance around the system gives

$$x_{\text{ac},F}F = Dx_{\text{ac},D} + Bx_{\text{ac},B}$$

$$0.08F = \left(825 \ \frac{\text{kmol}}{\text{h}}\right)(0.10) + 0 \ \frac{\text{kmol}}{\text{h}}$$

Solving for F gives

$$F = \frac{\left(825 \ \frac{\text{kmol}}{\text{h}}\right)(0.10) + 0 \ \frac{\text{kmol}}{\text{h}}}{0.08}$$

$$= 1031.25 \ \text{kmol/h}$$

An overall mole balance around the system gives

$$F = D + B$$

Rearranging to solve for B gives

$$B = F - D = 1031.25 \ \frac{\text{kmol}}{\text{h}} - 825 \ \frac{\text{kmol}}{\text{h}}$$

$$= 206.25 \ \text{kmol/h}$$

Because the distillate stream contains 76% acetic acid, the mole fraction of acetic acid in the distillate stream, $x_{\text{acid},D}$, is given as 0.76. Because the feed is 61.8% acetic acid, the mole fraction of acetic acid in the feed, $x_{\text{acid},F}$, is 0.618. The mole balance of acetic acid around the system gives

$$x_{\text{acid},F}F = x_{\text{acid},D}D + x_{\text{acid},B}B$$

Rearranging and solving,

$$x_{\text{acid},B} = \frac{x_{\text{acid},F}F - x_{\text{acid},D}D}{B}$$

$$= \frac{(0.618)\left(1031.25 \ \frac{\text{kmol}}{\text{h}}\right) - (0.76)\left(825 \ \frac{\text{kmol}}{\text{h}}\right)}{206.25 \ \frac{\text{kmol}}{\text{h}}}$$

$$= 0.050$$

The answer is (B).

Why Other Options Are Wrong

(A) This incorrect answer uses the mole fraction of acetic acid in the feed stream instead of the mole fraction of acetic acid in the distillate when performing the acetone mole balance.

(C) This incorrect answer uses the mole fraction of the acetic anhydride in the bottom stream instead of the mole fraction of acetone in the feed when performing the acetone mole balance.

(D) This incorrect answer uses the mole fraction of acetic acid in the feed instead of the mole fraction of acetone in the distillate when performing the acetone mole balance.

SOLUTION 75

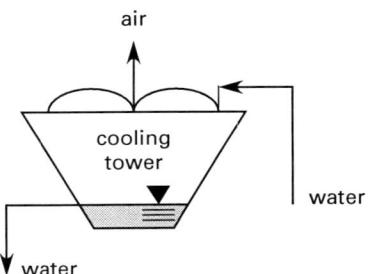

The ambient wet-bulb temperature, T_{bulb}, is given as 80.0°F. The hot water temperature, T_{w2}, is given as 104.0°F, and the cold water temperature, T_{w1}, is given as 89.0°F. The range is

$$R = T_{w2} - T_{w1} = 104.0°F - 89.0°F$$
$$= 15°F$$

To calculate the tower characteristic using the Chebyshev method, it is necessary to calculate four temperatures.

These four temperatures are calculated starting at the cold water temperature. The cold water temperature is increased by the following fractions of the range. The intervals are $0.1R$, $0.4R$, $0.6R$, and $0.9R$. The first temperature is

$$T_{w1} + 0.1R = 89.0°F + (0.1)(15°F)$$
$$= 90.50°F$$

The second temperature is

$$T_{w1} + 0.4R = 89.0°F + (0.4)(15°F)$$
$$= 95.0°F$$

The third temperature is

$$T_{w1} + 0.6R = 89.0°F + (0.6)(15°F)$$
$$= 98.0°F$$

The fourth temperature is

$$T_{w1} + 0.9R = 89.0°F + (0.9)(15°F)$$
$$= 102.5°F$$

Use these four temperatures to calculate the enthalpy difference between the enthalpy given by the water operating line and the enthalpy given by the air operating line.

At 90.50°F, the enthalpy from the water operating line is

$$h_w = -77.9921 \, \frac{\text{Btu}}{\text{lbm}} + \left(1.4877 \, \frac{\text{Btu}}{\text{lbm-°F}}\right)(90.50°F)$$
$$= 56.6448 \, \text{Btu/lbm}$$

At 90.50°F, the enthalpy from the air operating line is

$$h_a = -103.0606 \, \frac{\text{Btu}}{\text{lbm}} + \left(1.64863 \, \frac{\text{Btu}}{\text{lbm-°F}}\right)(90.50°F)$$
$$+ \left(2.96296 \times 10^{-6} \, \frac{\text{Btu}}{\text{lbm-°F}^2}\right)(90.50°F)^2$$
$$= 46.1647 \, \text{Btu/lbm}$$

A summary of the calculations of h_w, h_a, and $1/(h_w - h_a)$ is presented in the following table at selected temperatures of 90.50°F, 95.00°F, 98.00°F, and 102.50°F.

description	T_w (°F)	water side, h_w (Btu/lbm)	air side, h_a (Btu/lbm)	inverse enthalpy difference, $\frac{1}{h_w - h_a}$ (lbm/Btu)
$T_{w1} + 0.1R$	90.50	56.6448	46.1647	0.0954
$T_{w1} + 0.4R$	95.00	63.3394	53.5860	0.1025
$T_{w1} + 0.6R$	98.00	67.8025	58.5336	0.1079
$T_{w1} + 0.9R$	102.50	74.4972	65.9551	0.1171

From the Merkel equation, the tower characteristic is

$$\frac{KaV}{L} = \int_{T_{w1}}^{T_{w2}} \frac{dT}{h_w - h_a}$$

This integration is approximated by the Chebyshev method as

$$\frac{KaV}{L} = \left(\frac{R}{4}\right)\left(\frac{1}{\Delta h_1} + \frac{1}{\Delta h_2} + \frac{1}{\Delta h_3} + \frac{1}{\Delta h_4}\right)$$
$$= \left(\frac{15°F}{4}\right)\begin{pmatrix} 0.0954 \, \frac{\text{lbm}}{\text{Btu}} + 0.1025 \, \frac{\text{lbm}}{\text{Btu}} \\ + 0.1079 \, \frac{\text{lbm}}{\text{Btu}} + 0.1171 \, \frac{\text{lbm}}{\text{Btu}} \end{pmatrix}$$
$$= 1.586 \, \text{lbm-°F/Btu} \quad (1.6 \, \text{lbm-°F/Btu})$$

The answer is (C).

Why Other Options Are Wrong

(A) This incorrect answer calculates the range by subtracting the outlet air wet-bulb temperature from the hot water temperature.

(B) This incorrect answer omits the enthalpy difference at 102.5°F when calculating the tower characteristic.

(D) This answer fails to divide by the number of intervals when calculating the tower characteristic.

KINETICS

SOLUTION 76

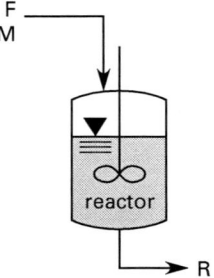

The initial reaction rate is

$$r_R = \frac{dC_R}{dt}$$

From the table, the reaction rate increases four times when the concentration of reactant F is doubled. The reaction rate increases nine times when the concentration of component F is tripled. The reaction is second-order with respect to reactant F. The reaction rate does not change at all if the concentration of component M is changed so the reaction is zero order with respect to reactant M.

For the first experiment, the temperature is given as

$$T_{25} = 25°C + 273° = 298K$$

From experiments 2–7, the temperature is given as

$$T_{30} = 30°C + 273° = 303K$$

At 25°C, the initial rate of formation of reactant R, r_{25}, is given as 1.5×10^{-3} mol/L·s. At 30°C, the initial rate of formation of reactant R, r_{30}, is given as 2.7×10^{-3} mol/L·s. Because the reaction is second order with respect to reactant F, the rate constant at 25°C is

$$k_{25} = \frac{r_{25}}{C_F^2}$$

At 30°C, the rate constant is

$$k_{30} = \frac{r_{30}}{C_F^2}$$

The ratio of the rate constants at 25°C and 30°C is

$$\frac{k_{25}}{k_{30}} = \frac{\frac{r_{25}}{C_F^2}}{\frac{r_{30}}{C_F^2}} = \frac{r_{25}}{r_{30}} = \frac{1.5 \times 10^{-3} \frac{\text{mol}}{\text{L·s}}}{2.7 \times 10^{-3} \frac{\text{mol}}{\text{L·s}}}$$
$$= 0.556$$

The universal gas constant, R, is 8.314×10^{-3} kJ/mol·K.

For the reaction given in the problem statement, the activation energy can be calculated as

$$k = k_0 e^{-E/RT}$$
$$\ln \frac{k_{25}}{k_{30}} = \left(\frac{E}{R}\right)\left(\frac{1}{T_{30}} - \frac{1}{T_{25}}\right)$$
$$E_a = \frac{T_{25} T_{30} R}{T_{25} - T_{30}} \ln \frac{k_{25}}{k_{30}}$$
$$= \frac{(298K)(303K)\left(8.314 \times 10^{-3} \frac{\text{kJ}}{\text{mol·K}}\right)}{298K - 303K}$$
$$\times \ln 0.556$$
$$= 88.13 \text{ kJ/mol} \quad (88 \text{ kJ/mol})$$

The answer is (A).

Why Other Options Are Wrong

(B) This incorrect answer does not use the universal gas constant when calculating the activation energy.

(C) This incorrect answer uses an incorrect universal gas constant (1.98 cal/mol·K) when calculating the activation energy.

(D) This answer uses an incorrect universal gas constant (10.73 ft³-lbm/in²-lbmol-°R) when calculating the activation energy.

SOLUTION 77

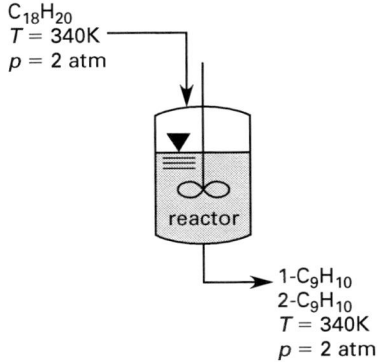

The reaction is

$$C_{18}H_{20}(l) \leftrightarrow C_9H_{10}(l) + C_9H_{10}(l)$$

Representing the component $C_{18}H_{20}$ with the letter M and the products of the reaction with the letters R and S, this reaction can be represented as

$$M(l) \leftrightarrow R(l) + S(l)$$

The concentration equilibrium constant at 340K, K_C, is given as 10 mol/L.

At equilibrium, the number of moles per liter of reactant M that reacted divided by the number of moles of M per liter initially present is X.

The initial concentration of the reactant M, C_{M0}, is given as 7.174 mol/L.

For the batch system, the stoichiometric table is as follows.

species	symbol	initial (mol/L)	change (mol/L)	remaining
$C_{18}H_{20}$	M	C_{M0}	$-C_{M0}X$	$C_{M0}(1-X)$
C_9H_{10}	R	0	$C_{M0}X$	$C_{M0}X$
C_9H_{10}	S	0	$C_{M0}X$	$C_{M0}X$
	total	C_{M0}		$C_{M0}(1+X)$

For this batch system, the concentration of the reactant M is

$$C_M = C_{M0}(1 - X)$$

The concentration of the product R is

$$C_R = C_{M0}X$$

The concentration of the product S is

$$C_S = C_{M0}X$$

At equilibrium, the concentration of the reactant M and the products R and S are related to the concentration equilibrium constant. The relationship is

$$K_C = \frac{C_R C_S}{C_M}$$

At equilibrium,

$$K_C = \frac{C_{M0}X C_{M0}X}{C_{M0}(1 - X)}$$
$$= \frac{C_{M0}X^2}{1 - X}$$
$$= 10 \text{ mol/L}$$

Solving for X gives

$$X = \left(\frac{1}{2C_{M0}}\right)\left(-K_C + \sqrt{K_C^2 + 4C_{M0}K_C}\right)$$
$$= 0.674 \quad (0.67)$$

The answer is (C).

Why Other Options Are Wrong

(A) This incorrect answer is consistent with the reasoning that because the reaction is reversible, the rate of the forward reaction equals the rate of the reverse reaction. In this case, the net change of concentration of any of the reagents would be zero.

(B) Instead of using the concentration equilibrium constant for the forward reaction, this incorrect answer uses the concentration equilibrium constant for the reverse reaction.

(D) The equilibrium constant equation was incorrectly set up, squaring the denominator and ignoring the lack of units on the right side of the equation.

SOLUTION 78

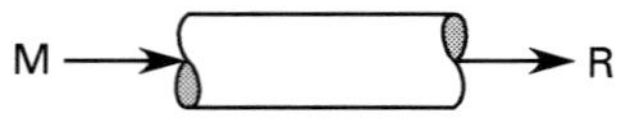

The specific reaction rate at 25°C, k, is given as 0.258 min^{-1}. The flow rate, Q_0, is given as 12 L/min. The entering molar flow rate, C_{M0}, is given as 1.5 mol/L. Because the conversion of M is given as 85%, the concentration of species M (assuming constant density) is

$$C_M = (1 - x)\,C_{M0}$$
$$= (1 - 0.85)\left(1.5\,\frac{\text{mol}}{\text{L}}\right)$$
$$= 0.225 \text{ mol/L}$$

For a tubular reactor, the mole balance of species M is

$$\frac{dF_M}{dV} = -r_M$$

The entering molar flow rate to the reactor is the product of the entering concentration of M and the entering volumetric flow rate. For constant density, $Q = Q_0$, so that at any point in the reactor

$$F_M = C_M Q_0$$

For a first-order reaction, the rate law is

$$-r_M = kC_M$$

Because the volumetric flow rate is constant, the mole balance of species M is

$$\frac{dF_M}{dV} = \frac{d(C_M Q_0)}{dV} = Q_0 \frac{dC_M}{dV}$$
$$= -r_M$$

Substituting for r_M gives

$$Q_0 \frac{dC_M}{dV} = -r_M$$
$$= -kC_M$$

Rearranging gives

$$\left(-\frac{Q_0}{k}\right)\frac{dC_M}{C_M} = dV$$

When the volume of the reaction mixture produced in the reactor is zero, the concentration of species M is 1.5 mol/L.

$$-\frac{Q_0}{k}\int_{C_{M0}}^{C_M}\frac{dC_M}{C_M} = \int_0^V dV$$

This equation gives

$$V = \left(\frac{Q_0}{k}\right)\ln\frac{C_{M0}}{C_M}$$

$$= \left(\frac{12\,\frac{L}{min}}{0.258\,min^{-1}}\right)\ln\frac{1.5\,\frac{mol}{L}}{0.225\,\frac{mol}{L}}$$

$$= 88.24\text{ L}\quad(90\text{ L})$$

The answer is (C).

Why Other Options Are Wrong

(A) This answer applies an incorrect formula to calculate the volume of the reactor. This answer divides the specific reaction by the volumetric flow rate instead of doing the inverse operation.

(B) This answer erroneously uses the conversion to calculate the concentration of species M.

(D) This incorrect answer fails to take the natural logarithm in the expression of the volume of the reactor.

SOLUTION 79

The specific reaction rate at 77°F, k, is given as 0.006 min^{-1}. The constant flow rate given as

$$Q_0 = \left(158.5\,\frac{gal}{hr}\right)\left(\frac{1\,hr}{60\,min}\right)$$

$$= 2.64\text{ gal/min}$$

The initial concentration of M, C_{M0}, is given as 3.785 lbmol/gal. Because the conversion is given as 90%, the concentration of species M (for constant density) is

$$C_M = (1-x)C_{M0}$$

$$= (1-0.90)\left(3.785\,\frac{lbmol}{gal}\right)$$

$$= 0.3785\text{ lbmol/gal}$$

For an isothermal continuous-flow tubular reactor, the mole balance on species M is

$$\frac{dF_M}{dV} = -r_M$$

Because the reaction is first order, the rate law is

$$-r_M = kC_M$$

Because the volumetric flow rate is constant, the mole balance on species M is

$$\frac{dF_M}{dV} = \frac{d(C_M Q_0)}{dV} = Q_0\frac{dC_M}{dV}$$

$$= -r_M$$

Substituting for r_M gives

$$Q_0\frac{dC_M}{dV} = -r_M$$

$$= -kC_M$$

Rearranging gives

$$\left(-\frac{Q_0}{k}\right)\frac{dC_M}{C_M} = dV$$

Integrating gives

$$-\frac{Q_0}{k}\int_{C_{M0}}^{C_M}\frac{dC_M}{C_M} = \int_0^V dV$$

This equation gives

$$V = \frac{Q_0}{k}\ln\frac{C_{M0}}{C_M}$$

$$= \frac{2.64\,\frac{gal}{min}}{0.006\,min^{-1}}\ln\frac{3.785\,\frac{lbmol}{gal}}{0.3785\,\frac{lbmol}{gal}}$$

$$= 1013.14\text{ gal}\quad(1000\text{ gal})$$

The answer is (B).

Why Other Options Are Wrong

(A) This answer uses the following incorrect expression when calculating the volume.

$$V = \frac{k}{Q_0}\ln\frac{C_{M0}}{C_M}$$

(C) This incorrect answer does not take the logarithm when calculating the volume.

(D) This incorrect answer omits the conversion factor needed to convert the volumetric feed rate from gallons per hour to gallons per minute.

SOLUTION 80

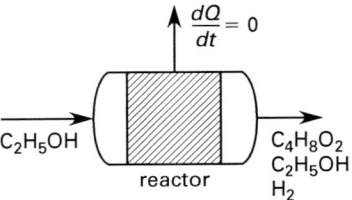

The basis of calculation is 100 mol/h of C_2H_5OH feed. The molar flow rate of ethanol entering the reactor, $F_{C_2H_5OH,in}$, is 100 mol/h. The ethanol conversion, X, is given as 0.35.

Because C_2H_5OH is a reactant, its stoichiometric coefficient is negative. From the balanced chemical reaction, the stoichiometric coefficient of C_2H_5OH, $\sigma_{C_2H_5OH}$, is -2. Because C_2H_5OH is disappearing, the rate of reaction of ethanol is negative. From the conversion definition,

$$-r_{C_2H_5OH} = X F_{C_2H_5OH,in}$$
$$= (0.35)\left(100\ \frac{\text{mol } C_2H_5OH}{\text{h}}\right)$$
$$= 35\ \text{mol } C_2H_5OH/\text{h}$$

The molar flow rate of ethanol out of the reactor is

$$F_{C_2H_5OH,out} = F_{C_2H_5OH,in} - r_{C_2H_5OH}$$
$$= 100\ \frac{\text{mol}}{\text{h}} - 35\ \frac{\text{mol}}{\text{h}}$$
$$= 65\ \text{mol/h}$$

The molar flow rate of ethyl acetate into the reactor, $F_{C_4H_8O_2,in}$, is given as 0 mol/h. From the balanced equation, the molar flow rate of $C_4H_8O_2$ out of the reactor is half the molar flow rate of the C_2H_5OH that reacts. The molar flow rate of ethyl acetate out of the reactor is

$$F_{C_4H_8O_2,out} = F_{C_4H_8O_2,in} + \frac{r_{C_2H_5OH}}{2}$$
$$= 0\ \frac{\text{mol}}{\text{h}} + \frac{35\ \frac{\text{mol}}{\text{h}}}{2}$$
$$= 17.5\ \text{mol/h}$$

The molar flow rate of hydrogen into the reactor, $F_{H_2,in}$, is given as 0 mol/h. The molar flow rate of hydrogen out of the reactor is

$$F_{H_2,out} = F_{H_2,in} + r_{C_2H_5OH}$$
$$= 0\ \frac{\text{mol}}{\text{h}} + 35\ \frac{\text{mol}}{\text{h}}$$
$$= 35\ \text{mol/h}$$

For the balanced reaction,

$$2C_2H_5OH \rightarrow C_2H_8O_2 + 2H_2$$

The standard heat of formation is given at a temperature, T_{ref}, of 298K. The temperature of the reaction is given as

$$T_R = 675°C + 273°$$
$$= 948K$$

The following table summarizes these results.

component	flow rate of reactants (mol/h)	flow rate out of the reactor (mol/h)	flow rate of products (mol/h)
C_2H_5OH	100	-35	65
$C_2H_8O_2$	0	17.5	17.5
H_2	0	35	35

The adiabatic energy balance for a flow system is

$$Q = 0 = \Delta H_R^0 + \Delta H_{298}^0 + \Delta H_P^0$$

The enthalpy difference to bring reactants from 948K to 298K is

$$\Delta H_R^0 = \left(\Sigma n_i C_{R,i}^0\right)(T_{ref} - T_R)$$
$$= \left(100\ \frac{\text{mol}}{\text{h}}\right)\left(26.7\ \frac{\text{cal}}{\text{mol·K}}\right)$$
$$\times (298K - 948K)$$
$$= -1\,735\,500\ \text{cal/h}$$

The enthalpy change for the reactants at 298K is

$$\Delta H_{298}^0 = \Sigma n_i \Delta H_{298,i}^0$$
$$= \left(-35\ \frac{\text{mol}}{\text{h}}\right)\left(-52\,230\ \frac{\text{cal}}{\text{mol}}\right)$$
$$+ \left(17.5\ \frac{\text{mol}}{\text{h}}\right)\left(-102\,020\ \frac{\text{cal}}{\text{mol}}\right)$$
$$+ \left(35\ \frac{\text{mol}}{\text{h}}\right)\left(0\ \frac{\text{cal}}{\text{mol}}\right)$$
$$= 42\,700\ \text{cal/h}$$

The enthalpy change to bring products from 298K to the temperature T is

$$\Delta H_P^0 = \left(\Sigma n_i C_{p,i}^0\right)(T - T_{ref})$$
$$= \left(\begin{array}{l}\left(65\ \frac{\text{mol}}{\text{h}}\right)\left(26.7\ \frac{\text{cal}}{\text{mol·K}}\right)\\ + \left(17.5\ \frac{\text{mol}}{\text{h}}\right)\left(9.6\ \frac{\text{cal}}{\text{mol·K}}\right)\\ + \left(35\ \frac{\text{mol}}{\text{h}}\right)\left(7\ \frac{\text{cal}}{\text{mol·K}}\right)\end{array}\right)$$
$$\times (T - 298K)$$
$$= \left(2148.5\ \frac{\text{cal}}{\text{h·K}}\right)(T - 298K)$$

The adiabatic energy balance is

$$Q = 0 = \Delta H_R^0 + \Delta H_{298}^0 + \Delta H_p^0$$
$$0 = -1\,735\,500\ \frac{\text{cal}}{\text{h}} + 42\,700\ \frac{\text{cal}}{\text{h}} + \left(2148.5\ \frac{\text{cal}}{\text{h·K}}\right)$$
$$\times (T - 298K)$$

Solving for the temperature gives

$$T = \frac{-\Delta H_R^0 - \Delta H_{298}^0 + (\Sigma n_i C_{p,i}^0) T_{\text{ref}}}{\Sigma n_i C_{p,i}^0}$$

$$= \frac{-\left(-1\,735\,500 \ \frac{\text{cal}}{\text{h}}\right) - 42\,700 \ \frac{\text{cal}}{\text{h}} + \left(2148.5 \ \frac{\text{cal}}{\text{h·K}}\right)(298\text{K})}{2148.5 \ \frac{\text{cal}}{\text{h·K}}} - 273°$$

$$= 812.9°\text{C} \quad (810°\text{C})$$

The answer is (B).

Why Other Options Are Wrong

(A) This incorrect answer omits the reactants' enthalpy change from 948K to 298K when calculating the temperature of the gases leaving the reactor.

(C) This incorrect answer omits the reactants' enthalpy change at 298K when calculating the temperature of the gases leaving the adiabatic reactor.

(D) This incorrect answer uses the opposite sign of the reactants' enthalpy change at 298K when calculating the temperature of the gases leaving the adiabatic reactor.

SOLUTION 81

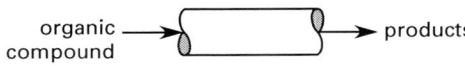

The temperature is given as

$$T = 720°\text{C} + 273° = 993\text{K}$$

The universal gas constant, R, is 0.082 atm·L/mol·K. The pressure, p, is given as 1 atm. The reaction constant, k, is given as 0.0182 mol/L·s·atm$^{1.5}$. The time, t, is given as 3 s. Because the organic compound is an ideal gas, its initial concentration is

$$C_{A0} = \frac{p}{RT} = \frac{1 \text{ atm}}{\left(0.082 \ \frac{\text{atm·L}}{\text{mol·K}}\right)(993\text{K})}$$

$$= 0.0123 \text{ mol/L}$$

The partial pressure of the organic compound is

$$p_A = C_A RT$$

Because the reaction order is 1.5, the reaction rate law is

$$\frac{-dC_A}{dt} = -r_A = k_1 (p_A)^{1.5} = k_1 \sqrt{(p_A)^3}$$

Replacing the partial pressure of the organic compound gives

$$\frac{-dC_A}{dt} = k_1 \sqrt{(C_A RT)^3}$$

Rearranging gives

$$\frac{-dC_A}{\sqrt{(C_A)^3}} = k_1 \sqrt{(RT)^3} dt$$

Integrating both sides of the preceding equation gives

$$\int_{C_{A0}}^{C_A} \frac{-dC_A}{\sqrt{(C_A)^3}} = \int_0^t k_1 \sqrt{(RT)^3} dt$$

C_{A0}, k_1, R, and T are constants.

$$-\left(\frac{\frac{1}{\sqrt{C_A}}}{-0.5} - \frac{\frac{1}{\sqrt{C_{A0}}}}{-0.5}\right) = k_1 \sqrt{(RT)^3} t$$

Rearranging and solving for C_A gives

$$C_A = \frac{1}{\left(\frac{k_1 t \sqrt{(RT)^3}}{2} + \frac{1}{\sqrt{C_{A0}}}\right)^2}$$

$$= \frac{1}{\left(\frac{\left(0.0182 \ \frac{\text{mol}}{\text{L·s·atm}^{1.5}}\right)(3 \text{ s}) \times \sqrt{\left(\left(0.082 \ \frac{\text{atm·L}}{\text{mol·K}}\right)(993\text{K})\right)^3}}{2} + \frac{1}{\sqrt{0.0123 \ \frac{\text{mol}}{\text{L}}}}\right)^2}$$

$$= 0.00118 \text{ mol/L} \quad (1.2 \times 10^{-3} \text{ mol/L})$$

The answer is (A).

Why Other Options Are Wrong

(B) This incorrect answer omits the exponent on the RT term when integrating.

(C) This incorrect answer omits the exponent on the concentration of the organic compound when calculating the integration.

(D) This incorrect answer integrates the reaction rate incorrectly, excluding the exponent on the first term of the integration. This answer adds terms in different units.

SOLUTION 82

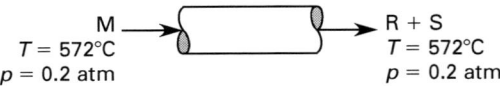

The stoichiometric table is

component	initial concentration (kmol/h)	concentration (kmol/h)
M	C_{M0}	$C_{M0}(1-X)$
R	0	$C_{M0}X$
S	0	$C_{M0}X$
total		$C_{M0}(1+X)$

The gas-phase partial pressure of component M equals the concentration of component M over the total concentration times the total pressure. The gas-phase partial pressure of component M is

$$p_M = \left(\frac{C_{M0}(1-X)}{C_{M0}(1+X)}\right)p = \left(\frac{C_{M0}(1-X)}{C_{M0}(1+X)}\right)(0.2 \text{ atm})$$
$$= \left(\frac{1-X}{1+X}\right)(0.2 \text{ atm})$$

The gas-phase partial pressure of component R equals the concentration of component R over the total concentration times the total pressure. The gas-phase partial pressure of component R is

$$p_R = \left(\frac{C_{M0}X}{C_{M0}(1+X)}\right)p = \left(\frac{X}{1+X}\right)p$$
$$= \left(\frac{X}{1+X}\right)(0.2 \text{ atm})$$

The gas-phase partial pressure of component S equals the concentration of component S over the total concentration times the total pressure. The gas-phase partial pressure of component S is

$$p_S = \left(\frac{C_{M0}X}{C_{M0}(1+X)}\right)p = \left(\frac{X}{1+X}\right)(0.2 \text{ atm})$$

Substitution of p_R and p_S in the expression given for p_{RS} in the problem statement gives

$$p_{RS} = \frac{p_R + p_S}{2}$$
$$= \frac{\left(\frac{X}{1+X}\right)(0.2 \text{ atm}) + \left(\frac{X}{1+X}\right)(0.2 \text{ atm})}{2}$$
$$= \left(\frac{X}{1+X}\right)(0.2 \text{ atm})$$

Substitution of the values of p_M, p_R, p_S, p_{RS}, and p in the expression for r_M given in the problem statement yields

$$r_M = \frac{\left(0.0697 \frac{\frac{\text{kmol M converted}}{\text{h}}}{\text{kg of catalyst·atm}}\right)\left(\left(\frac{1-X}{1+X}\right)(0.2 \text{ atm}) - \frac{\left(\left(\frac{X}{1+X}\right)(0.2 \text{ atm})\right)\times\left(\left(\frac{X}{1+X}\right)(0.2 \text{ atm})\right)}{0.203 \text{ atm}}\right)}{\left(1 + (0.445 \text{ atm}^{-1})\left(\frac{1-X}{1+X}\right)(0.2 \text{ atm}) + (1.526 \text{ atm}^{-1})\left(\frac{X}{1+X}\right)(0.2 \text{ atm})\right)^2}$$

Rearranging and taking the inverse gives

$$\frac{1}{r_M} = \frac{\left(1 + (0.445 \text{ atm}^{-1})(0.2 \text{ atm})\left(\frac{1-X}{1+X}\right) + (1.526 \text{ atm}^{-1})(0.2 \text{ atm})\left(\frac{X}{1+X}\right)\right)^2}{\left(0.0697 \frac{\frac{\text{kmol M converted}}{\text{h}}}{\text{kg of catalyst·atm}}\right) \times \left((0.2 \text{ atm})\left(\frac{1-X}{1+X}\right) - \frac{(0.2 \text{ atm})^2 X^2}{(0.203 \text{ atm})(1+X)^2}\right)}$$

The result from the following integration equals the number of kilograms of catalyst divided by the feed rate of component A.

$$\frac{m_{\text{catalyst}}}{F_{M0}} = \int \frac{dX}{r_M}$$

$$= \int_0^{0.2} \frac{\left(1 + (0.445 \text{ atm}^{-1})(0.2 \text{ atm})\left(\frac{1-X}{1+X}\right) + (1.526 \text{ atm}^{-1})(0.2 \text{ atm})\left(\frac{X}{1+X}\right)\right)^2}{\left(0.0697 \frac{\frac{\text{kmol M converted}}{\text{h}}}{\text{kg of catalyst·atm}}\right) \times \left((0.2 \text{ atm})\left(\frac{1-X}{1+X}\right) - \frac{(0.2 \text{ atm})^2 X^2}{(0.203 \text{ atm})(1+X)^2}\right)} \times dX$$

The limits of the integration are the mole fraction of component M converted, which is 0 at time zero, and the final mole fraction of component M converted, which is given as 0.2. This integration gives a value of 21.744 kg catalyst/kmol of component M per hour in the feed.

$$\int \frac{dX}{r_M} = 21.744 \text{ kg of catalyst·h/kmol M}$$

The feed rate of component A, F_{M0}, is given as 20 kmol M/h.

$$\int \frac{dX}{r_M} = \frac{m_{\text{catalyst}}}{F_{M0}}$$
$$= 21.744 \text{ kg of catalyst·h/kmol M}$$

Solving for the number of kilograms of catalyst gives

$$m_{\text{catalyst}} = \left(21.744 \ \frac{\text{kg of catalyst}}{\frac{\text{kmol M}}{\text{h}}}\right)\left(20 \ \frac{\text{kmol M}}{\text{h}}\right)$$
$$= 434.9 \text{ kg of catalyst} \quad (430 \text{ kg})$$

The answer is (D).

Alternate Solution

The integration can also be done using Simpson's approximation rule as follows.

$$\int_a^b f(X)dX = \left(\frac{\Delta X}{3}\right)(f_o + 4f_1 + 2f_2 + \cdots$$
$$+ f_n + 4f_{n-1} + 2f_{n-2})$$

With n even,

X	$1/r_M$ (h·kg of catalyst/ kmol)	partial sum (kg of catalyst·h/ kg M converted)
0	85.07	85.07
0.02	88.99	355.9
0.04	93.14	186.3
0.06	97.55	390.2
0.08	102.3	204.5
0.10	107.3	429.1
0.12	112.6	225.2
0.14	118.4	473.4
0.16	124.5	249.0
0.18	131.2	524.6
0.2	138.3	138.3
	total	3262.0

The area under the integration is

$$A = \left(\frac{\Delta X}{3}\right)(\text{total})$$
$$= \left(\frac{0.02}{3}\right)\left(3262 \ \frac{\frac{\text{kg of catalyst}}{\text{kg M converted}}}{\text{h}}\right)$$
$$= 21.747 \text{ kg of catalyst·h/kg M converted}$$

$$\mu_{\text{catalyst}} = \left(21.747 \ \frac{\frac{\text{kg of catalyst}}{\text{kmol M converted}}}{\text{h}}\right)$$
$$\times \left(20 \ \frac{\text{kmol M}}{\text{h}}\right)$$
$$= 434.9 \text{ kg of catalyst} \quad (430 \text{ kg})$$

Why Other Options Are Wrong

(A) This incorrect answer integrates the inverse of the function.

(B) This incorrect answer omits the rate constant C in the function when performing the integration.

(C) This incorrect answer omits the exponent two in the numerator of the function when performing the integration.

SOLUTION 83

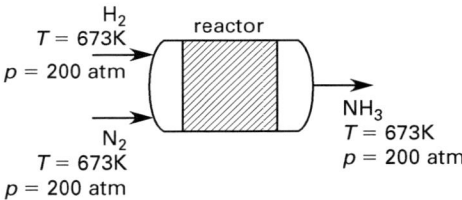

The initial concentration of N_2, $C_{N_2,\text{initial}}$, is given as 10 mol/L. The initial concentration of H_2, $C_{H_2,\text{initial}}$, is given as 30 mol/L. The equilibrium concentration of N_2, $C_{N_2,e}$, is given as 0.5 mol/L. The initial concentration of NH_3, $C_{NH_3,\text{initial}}$, is 0 mol/L.

The number of moles of N_2 reacted per liter is

$$C_{N_2,\text{reacted}} = C_{N_2,\text{initial}} - C_{N_2,e}$$
$$= 10 \ \frac{\text{mol}}{\text{L}} - 0.5 \ \frac{\text{mol}}{\text{L}}$$
$$= 9.5 \text{ mol/L}$$

From the stoichiometry of the reaction, the number of moles of H_2 reacted per liter is

$$C_{H_2,\text{reacted}} = 3C_{N_2,\text{reacted}}$$
$$= (3)\left(9.5 \ \frac{\text{mol}}{\text{L}}\right)$$
$$= 28.5 \text{ mol/L}$$

The equilibrium concentration of H_2 is

$$C_{H_2,e} = C_{H_2,\text{initial}} - C_{H_2,\text{reacted}}$$
$$= 30 \; \frac{\text{mol}}{\text{L}} - 28.5 \; \frac{\text{mol}}{\text{L}}$$
$$= 1.5 \; \text{mol/L}$$

From the stoichiometry of the reaction, the number of moles of NH_3 formed per liter is

$$C_{NH_3,\text{formed}} = 2C_{N_2,\text{reacted}}$$
$$= (2)\left(9.5 \; \frac{\text{mol}}{\text{L}}\right)$$
$$= 19.0 \; \text{mol/L}$$

The equilibrium concentration of NH_3 is

$$C_{NH_3,e} = C_{NH_3,\text{initial}} + C_{NH_3,\text{formed}}$$
$$= 0 \; \frac{\text{mol}}{\text{L}} + 19.0 \; \frac{\text{mol}}{\text{L}}$$
$$= 19.0 \; \text{mol/L}$$

At equilibrium, the concentrations of the reacting species are correlated by the following relationship.

$$K_C = \frac{(C_{NH_3,e})^2}{C_{N_2,e}(C_{H_2,e})^3}$$
$$= \frac{\left(19.0 \; \frac{\text{mol}}{\text{L}}\right)^2}{\left(0.5 \; \frac{\text{mol}}{\text{L}}\right)\left(1.5 \; \frac{\text{mol}}{\text{L}}\right)^3}$$
$$= 213.93 \; \text{L}^2/\text{mol}^2 \quad (210 \; \text{L}^2/\text{mol}^2)$$

The answer is (C).

Why Other Options Are Wrong

(A) This incorrect answer calculates the inverse of the equilibrium concentration constant.

(B) This answer incorrectly omits the exponent on the concentration of NH_3 when calculating the equilibrium concentration constant.

(D) This answer incorrectly omits the exponent on the concentration of H_2 when calculating the equilibrium concentration constant.

SOLUTION 84

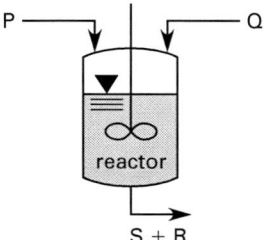

The initial concentration of P, C_{P0}, is given as 0.368 mol/L. The initial concentration of Q, C_{Q0}, is given as 0.785 mol/L. Because the stoichiometry of the reaction is of the form $P + Q \rightarrow R + S$, assume a second-order reaction. For the second-order reaction, the ratio of the initial concentration of Q over the initial concentration of P is

$$M = \frac{C_{Q0}}{C_{P0}} = \frac{0.785 \; \frac{\text{mol}}{\text{L}}}{0.368 \; \frac{\text{mol}}{\text{L}}}$$
$$= 2.133$$

From the stoichiometry of the reaction, the conversion of P is

$$X_P = \frac{C_{P0} - C_P}{C_{P0}}$$

Substituting,

$$X_P = \frac{C_{P0} - (C_{P0} - C_R)}{C_{P0}} = \frac{C_R}{C_{P0}}$$
$$= \frac{C_R}{0.368 \; \frac{\text{mol}}{\text{L}}}$$

For instance, at time 4 min, the value of C_R is given as 0.06 mol/L. Replacing the value of C_R in the preceding equation,

$$X_P = \frac{C_R}{0.368 \; \frac{\text{mol}}{\text{L}}} = \frac{0.06 \; \frac{\text{mol}}{\text{L}}}{0.368 \; \frac{\text{mol}}{\text{L}}}$$
$$= 0.163$$

Replacing this value and the 2.133 value of M in the following equation,

$$\ln \frac{M - X_P}{M(1 - X_P)} = (C_{Q0} - C_{P0})kt$$
$$\ln \frac{2.133 - 0.163}{(2.133)(1 - 0.163)} = (C_{Q0} - C_{P0})kt$$
$$0.098435 = (C_{Q0} - C_{P0})kt$$

This value appears in the fourth column of the following table. Similarly, calculate the values for the other times. Assuming that a plot of the fourth column as ordinate versus time is a straight line, the slope of this line, m, may be calculated. For instance, considering the first two rows of the data given in the problem statement, the slope is

$$m = \frac{0.098435 - 0}{4 \; \text{min} - 0 \; \text{min}} = 0.024609$$

Similarly, consider the two consecutive rows of the data given to calculate the slope. The results are shown in the last column of the following table. Because the reaction is P + Q → R + S, assume a second-order reaction.

time (min)	C_R (mol/L)	X_P	$\ln\dfrac{M - X_P}{M(1 - X_P)}$	slope, m
0	0	0	0	
4	0.06	0.163	0.098435	$\dfrac{0.098435 - 0}{4 \text{ min} - 0 \text{ min}}$ $= 0.024609$
10	0.135	0.367	0.268473	$\dfrac{0.268473 - 0.098435}{10 \text{ min} - 4 \text{ min}}$ $= 0.028340$
16	0.188	0.511	0.441523	$\dfrac{0.441523 - 0.268473}{16 \text{ min} - 10 \text{ min}}$ $= 0.028842$
20	0.218	0.592	0.571390	$\dfrac{0.571390 - 0.441523}{20 \text{ min} - 16 \text{ min}}$ $= 0.032467$
26	0.248	0.674	0.741080	$\dfrac{0.741080 - 0.571390}{26 \text{ min} - 20 \text{ min}}$ $= 0.028282$
			sum	0.142540
			average slope	$\dfrac{\text{sum}}{5} = \dfrac{0.142540}{5}$ $= 0.0285080$

A straight-line plot of the natural logarithm term versus the time in minutes confirms that the reaction is second-order.

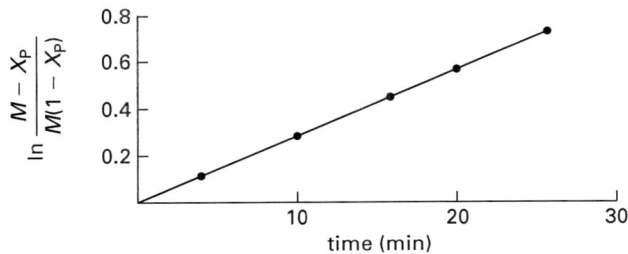

The slope of the straight line, m, is $0.0285080 \text{ min}^{-1}$. The reaction rate constant is

$$k = \frac{m}{C_{Q0} - C_{P0}}$$

$$= \frac{0.0285080 \text{ min}^{-1}}{0.785 \dfrac{\text{mol}}{\text{L}} - 0.368 \dfrac{\text{mol}}{\text{L}}}$$

$$= 0.06836 \text{ L/mol·min}$$

The rate law is

$$-r_P = kC_P C_Q$$

$$= \left(0.06836 \dfrac{\text{L}}{\text{mol·min}}\right) C_P C_Q$$

$$\left(\left(0.07 \dfrac{\text{L}}{\text{mol·min}}\right) C_P C_Q\right)$$

The answer is (B).

Why Other Options Are Wrong

(A) This incorrect answer uses the slope of the straight line of the second-order reaction in the rate law instead of the rate constant.

(C) This incorrect answer uses the slope of the straight line of the second-order reaction in the rate law instead of the rate constant, and it excludes the contribution of the concentration of the product Q to the rate law.

(D) This incorrect answer uses the concentration of P only when determining the rate law. This answer does not consider the contribution of the product Q to the rate law.

SOLUTION 85

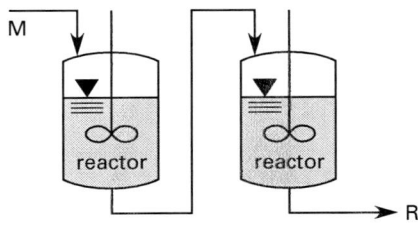

The initial concentration of M, C_0, is given as 1.0 mol/L. The rate constant, k, is given as 0.41 min^{-1}. The molar flow rate of M, F_0, is given as 109 mol/min. The overall production of R, F_R, is given as 87 mol/min. The fractional outlet conversion is

$$X_2 = \frac{F_R}{F_0} = \frac{87 \dfrac{\text{mol}}{\text{min}}}{109 \dfrac{\text{mol}}{\text{min}}} = 0.798$$

For the CSTRs in series, the material balance around either of the CSTRs is

$$\tau_i = \frac{C_o V_i}{F_o} = \frac{V_i}{v} = \frac{C_o(X_i - X_{i-1})}{-r_{A_i}}$$

The conversion definition for constant density is

$$X_i = \frac{C_o - C_i}{C_o}$$

For a first-order reaction,

$$\frac{V_i}{v} = \frac{C_o(X_i - X_{i-1})}{kC_i} = \frac{X_i - X_{i-1}}{k(1 - X_i)}$$

The volume for either of the CSTRs is

$$V_i = \left(\frac{v}{k}\right)\left(\frac{X_i - X_{i-1}}{1 - X_i}\right)$$

The total volume of both reactors is

$$V_1 + V_2 = \left(\frac{v}{k}\right)\left(\frac{X_1 - X_0}{1 - X_1} + \frac{X_2 - X_1}{1 - X_2}\right)$$

Minimizing the volume with respect to X_1,

$$0 = \frac{d}{dX_1}(V_1 + V_2)$$
$$= \left(\frac{v}{k}\right)\left(\frac{d}{dX_1}\left(\frac{X_1 - X_0}{1 - X_1}\right) + \frac{d}{dX_1}\left(\frac{X_2 - X_1}{1 - X_2}\right)\right)$$
$$= \left(\frac{v}{k}\right)\left(\frac{-1 + X_0}{(-1 + X_1)^2} + \frac{1}{-1 + X_2}\right)$$

Alternately,

$$\frac{1 - X_0}{(1 - X_1)^2} = \frac{1}{1 - X_2}$$

Rearranging,

$$\frac{1 - X_0}{(1 - X_1)^2} - \frac{1 - X_1}{(1 - X_1)^2} = \frac{1}{1 - X_2} - \frac{1}{1 - X_1}$$

$$\frac{X_1 - X_0}{(1 - X_1)^2} = \frac{1}{1 - X_2} - \frac{1}{1 - X_1}$$

In this problem, the inlet conversion is zero.

$$X_0 = 0$$

Note that for first-order CSTRs in series, minimum volume occurs when the volumes are equal, so $V_1 = V_2$.

Because the reaction is first order, the minimum reactor volume is achieved when the fractional conversion in the first reactor and the fractional conversion from the system satisfy the following equation.

$$\frac{X_1}{(1 - X_1)^2} = \frac{1}{1 - X_2} - \frac{1}{1 - X_1}$$

Solving for X_1 gives

$$X_1 = 1 - \sqrt{(1 - X_0)(1 - X_2)}$$
$$= 1 - \sqrt{(1 - 0)(1 - 0.798)}$$
$$= 0.551$$

The reaction rate in the first reactor is

$$-r_1 = kC_0(1 - X_1)$$
$$= (0.41 \text{ min}^{-1})\left(1 \frac{\text{mol}}{\text{L}}\right)(1 - 0.551)$$
$$= 0.184 \text{ mol/L·min}$$

The reaction rate in the second reactor is

$$-r_2 = kC_0(1 - X_2)$$
$$= (0.41 \text{ min}^{-1})\left(1 \frac{\text{mol}}{\text{L}}\right)(1 - 0.798)$$
$$= 0.083 \text{ mol/L·min}$$

For a first-order reaction in a CSTR, the total reactor volume is

$$V_1 + V_2 = F_0\left(\frac{X_1}{-r_1}\right) + F_0\left(\frac{X_2 - X_1}{-r_2}\right)$$
$$= \left(109 \frac{\text{mol}}{\text{min}}\right)\left(\frac{0.551}{0.184 \frac{\text{mol}}{\text{L·min}}}\right)$$
$$\quad + \left(109 \frac{\text{mol}}{\text{min}}\right)\left(\frac{0.798 - 0.551}{0.083 \frac{\text{mol}}{\text{L·min}}}\right)$$
$$= 651 \text{ L} \quad (650 \text{ L})$$

The answer is (C).

Why Other Options Are Wrong

(A) This incorrect answer reports the volume of one reactor instead of the total volume of the system of two reactors.

(B) This incorrect answer uses the production rate of R instead of the feed flow rate of M when calculating the minimum total volume.

(D) This answer erroneously uses the value of the reaction rate in the second reactor in place of the reaction rate in the first reactor when calculating the minimum total volume.

PLANT DESIGN AND OPERATION

SOLUTION 86

When designing a reactor, one must use materials that will transfer heat well, resist corrosion, and be strong enough to withstand the stresses of repeated and prolonged exposure to operating temperatures, pressure, vibration, and so on.

All the materials mentioned are corrosion resistant under the operating conditions. Any of the materials would provide enough heat conductivity. Because the coils are long, the material must be strong and not too soft. This eliminates copper. Silica does not perform well under tension. Both silica and nickel-free iron are brittle. Long coils made from these materials would be fragile. Fragility makes assembly and support difficult. Leaky coils can be disastrous considering that acetone is extremely flammable and its vapors may become explosive when mixed with air. A high-chromium-iron alloy is the best choice.

The answer is (B).

Why Other Options Are Wrong

(A) This incorrect answer selects a material that is fragile.

(C) This incorrect answer selects a material that is fragile.

(D) This incorrect answer selects a material that is fragile.

SOLUTION 87

The mass flow rate of the feed, $\dot{m}_F$, is given as 26.54 lbm/hr. The density, ρ, of R is given as 0.0453 lbm/ft^3. The volumetric flow rate of the feed is

$$\dot{V} = \frac{\dot{m}_F}{\rho} = \frac{26.54 \ \frac{\text{lbm}}{\text{hr}}}{0.0453 \ \frac{\text{lbm}}{\text{ft}^3}}$$

$$= 585.87 \ \text{ft}^3/\text{hr}$$

The conversion of R is given as

$$X = 0.1916 \ln z - 0.0865$$

Replacing for a conversion of 0.12 as given in the problem statement and solving for z gives

$$z = e^{\frac{X + 0.0865}{0.1916}} = e^{\frac{0.12 + 0.0865}{0.1916}}$$

$$= 2.938 \ \text{ft}^3\text{-hr/lbm}$$

The ratio of the volume of the reactor to the mass flow rate of the feed is

$$z = \frac{V}{\dot{m}_F} = 2.938 \ \text{ft}^3\text{-hr/lbm}$$

Solving for the volume gives

$$V = \dot{m}_F z = \left(26.54 \ \frac{\text{lbm}}{\text{hr}}\right)\left(2.938 \ \frac{\text{ft}^3\text{-hr}}{\text{lbm}}\right)$$

$$= 77.975 \ \text{ft}^3$$

The mean residence time is

$$\tau = \frac{V}{\dot{V}} = \frac{77.975 \ \text{ft}^3}{585.87 \ \frac{\text{ft}^3}{\text{hr}}}$$

$$= 0.133 \ \text{hr} \quad (0.1 \ \text{hr})$$

The answer is (C).

Why Other Options Are Wrong

(A) This answer erroneously calculates the volume of the material in the reactor using $V = \dot{m}_F/z$.

(B) This incorrect answer fails to take the natural logarithm in the correlation of the conversion of R versus z when calculating the ratio z.

(D) This incorrect answer exchanges the volume of the reactor with the volumetric flow rate of the feed when calculating the mean residence time.

SOLUTION 88

Draw a vertical line at a mole fraction of N equal to 0.2. This is line f. Draw a horizontal line at 800°C. This is line T. Lines f and T intercept at P$_1$. Line T intercepts two curves on the phase diagram. One intercept is at the point with coordinates (0.10, 800°C); the other intercept is at the point with coordinates (0.28, 800°C). (See the following illustration.)

The length of line T is

$$\overline{ab} = 0.28 - 0.10$$

$$= 0.18$$

The coordinates of point P$_1$ are (0.2, 800°C). The length of the segment from point P$_1$ to the intersection with the liquid-phase equilibrium curve is

$$\overline{P_1 b} = 0.28 - 0.2$$

$$= 0.08$$

Applying the lever rule gives

$$X = \frac{\overline{P_1 b}}{\overline{ab}} = \frac{0.08}{0.18}$$

$$= 0.44 \quad (0.4)$$

The answer is (C).

Illustration for Solution 88

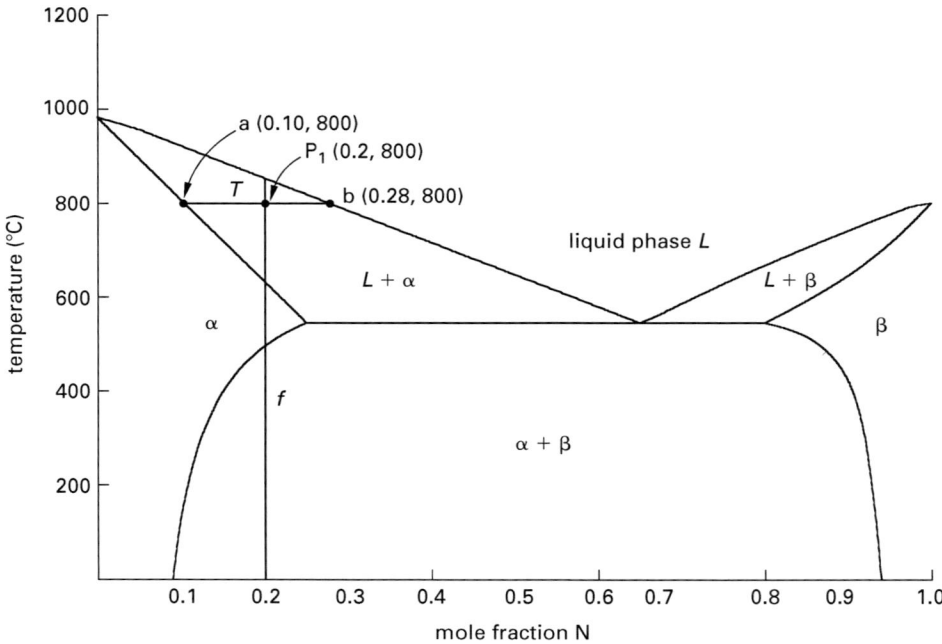

Why Other Options Are Wrong

(A) This incorrect answer gives the intersection of line T with the equilibrium curve on the left side of the phase diagram (i.e., the mole fraction of N at point a).

(B) This incorrect answer is the interception of the line T with the liquid-phase curve (i.e., the mole fraction of N at point b).

(D) This incorrect answer uses the segment on line T to the left of point P_1 instead of the segment on the right side of point P_1 when calculating the mole fraction.

$$\overline{aP_1} = 0.2 - 0.10 = 0.10$$

$$X = \frac{\overline{aP_1}}{\overline{ab}} = \frac{0.10}{0.18}$$

$$= 0.56$$

SOLUTION 89

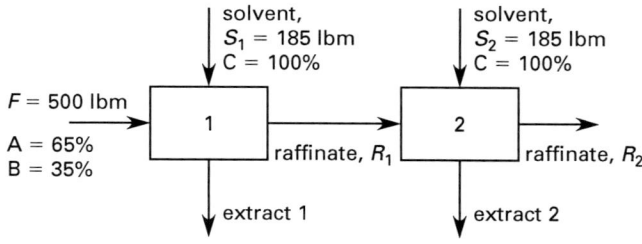

Because the feed contains 65% of A, the fraction of component A in the feed, $x_{A,F}$, is 0.65. Because the feed contains 35% of B, the fraction of component B in the feed, $x_{B,F}$, is 0.35. The total mass of A in the feed is

$$A_F = x_{A,F} F$$
$$= (0.65)(500 \text{ lbm})$$
$$= 325 \text{ lbm}$$

The total mass of B in the feed is

$$B_F = F - A_F$$
$$= 500 \text{ lbm} - 325 \text{ lbm}$$
$$= 175 \text{ lbm}$$

The total mass balance around extractor 1 gives the total mixture mass in extractor 1. The total mixture mass in extractor 1 is

$$M_1 = F + S_1$$
$$= 500 \text{ lbm} + 185 \text{ lbm}$$
$$= 685 \text{ lbm}$$

The mass fraction of A in M_1 is

$$x_{A,M_1} = \frac{A_F}{M_1} = \frac{325 \text{ lbm}}{685 \text{ lbm}}$$
$$= 0.4745 \quad (0.47)$$

The mass fraction of B in M_1 is

$$x_{B,M_1} = \frac{B_F}{M_1}$$
$$= \frac{175 \text{ lbm}}{685 \text{ lbm}}$$
$$= 0.2555 \quad (0.26)$$

The mass balance of C around stage 1 gives the mass fraction of C in M_1.

$$x_{C,M_1} = 1 - x_{A,M_1} - x_{B,M_1}$$
$$= 1 - 0.4745 - 0.2555$$
$$= 0.2700$$

Locate the feed point, F', with the coordinates (65 A, 35 B, 0 C) on the $\overline{AB}$ line.

The coordinates of the addition point, M_1', are (47.45 A, 25.55 B, 27.01 C). This point is located on the line that connects point F' with point C. By interpolation and with the help of the equilibrium diagram, locate the tie line passing through M_1'. This line intercepts the bimodal curve in R_1' and E_1'. From the diagram, the coordinates of point R_1' are

$$x_{A,R_1} = 0.36$$
$$x_{B,R_1} = 0.59$$

The mass balance around R_1 gives the mass fraction of C in R_1.

$$x_{C,R_1} = 1 - x_{A,R_1} - x_{B,R_1}$$
$$= 1 - 0.36 - 0.59$$
$$= 0.05$$

From the diagram, the coordinates of point E_1' are

$$x_{A,E_1} = 0.52$$
$$x_{B,E_1} = 0.12$$

The mass balance around E_1 gives the mass fraction of C in E_1.

$$x_{C,E_1} = 1 - 0.52 - 0.12$$
$$= 0.36$$

From point R_1' draw a line to point C. The total mass balance around stage 1 gives

$$R_1 + E_1 = M_1$$
$$= 685 \text{ lbm}$$
$$= F + S_1$$

Illustration for Solution 89

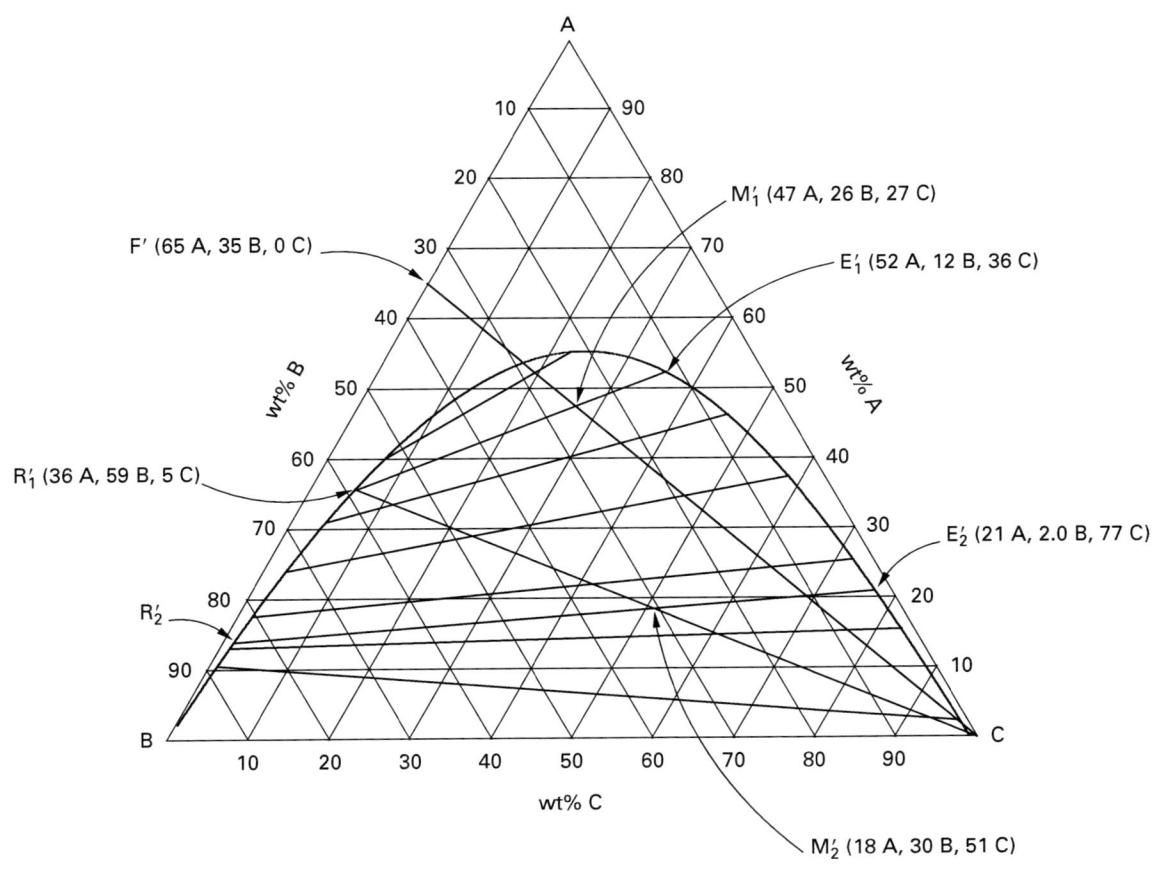

The mass balance of A around extractor 1 is

$$S_1 x_{A,S_1} + x_{A,F} F = x_{A,R_1} R_1 + x_{A,E_1} E_1$$
$$(185 \text{ lbm})(0) + (0.65)(500 \text{ lbm}) = 0.36 R_1 + 0.52 E_1$$

Solving the two previous equations simultaneously for R_1 and E_1 gives

$$R_1 = 195 \text{ lbm}$$
$$E_1 = 490 \text{ lbm}$$

The mass of A in R_1 is

$$A_{M_2} = x_{A,R_1} R_1 = (0.36)(195 \text{ lbm})$$
$$= 70.20 \text{ lbm}$$

The mass of B in R_1 is

$$B_{M_2} = x_{B,R_1} R_1 = (0.59)(195 \text{ lbm})$$
$$= 115.05 \text{ lbm}$$

The total mass balance around extractor 2 gives the total mixture mass in extractor 2.

$$M_2 = R_1 + S_2 = 195 \text{ lbm} + 185 \text{ lbm}$$
$$= 380 \text{ lbm}$$

The mass fraction of A in M_2 is

$$x_{A,M_2} = \frac{A_{M_2}}{M_2} = \frac{70.20 \text{ lbm}}{380 \text{ lbm}}$$
$$= 0.1847 \quad (0.18)$$

The mass fraction of B in M_2 is

$$x_{B,M_2} = \frac{B_{M_2}}{M_2} = \frac{115.05 \text{ lbm}}{380 \text{ lbm}}$$
$$= 0.3028 \quad (0.30)$$

The mass balance of C around stage 2 gives the mass fraction of C in M_2. The mass fraction of C in M_2 is

$$x_{C,M_2} = 1 - x_{A,M_2} - x_{B,M_2}$$
$$= 1 - 0.1847 - 0.3028$$
$$= 0.5125 \quad (0.51)$$

The coordinates of the addition point M_2' are (18 A, 30 B, 51 C). This point is located on the line that connects point R_1' with point C.

By interpolation and with the help of the equilibrium diagram, locate the tie line passing through M_2'. This line intercepts the bimodal curve in R_2' and E_2'.

From the diagram, the mass fraction of A in E_2, x_{A,E_2}, is 0.21.

The answer is (C).

Why Other Options Are Wrong

(A) This incorrect answer is the mass fraction of A in the raffinate produced in the second stage instead of the mass fraction of A in the extract produced in the second stage. This value is read from the diagram.

(B) This incorrect answer is the mass fraction of A in M_2 instead of the mass fraction of A in the extract produced in the second stage. This value is read from the diagram.

(D) This incorrect answer is the mass fraction of A in the extract produced in the first stage instead of the mass fraction of A in the extract produced in the second stage. This value is read from the diagram.

SOLUTION 90

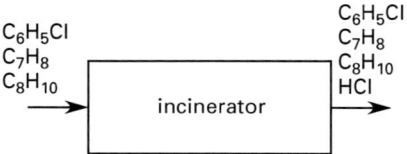

The DRE for the C_6H_5Cl is

$$DRE_{C_6} = \frac{\dot{m}_{in} - \dot{m}_{out}}{\dot{m}_{in}} \times 100\%$$

$$= \frac{154 \frac{\text{kg}}{\text{h}} - 0.012 \frac{\text{kg}}{\text{h}}}{154 \frac{\text{kg}}{\text{h}}} \times 100\%$$

$$= 99.9922\%$$

Because the DRE for the C_6H_5Cl is at least 99.99%, the C_6H_5Cl is in compliance. The DRE for the C_7H_8 is

$$DRE_{C_7} = \frac{\dot{m}_{in} - \dot{m}_{out}}{\dot{m}_{in}} \times 100\%$$

$$= \frac{431 \frac{\text{kg}}{\text{h}} - 0.036 \frac{\text{kg}}{\text{h}}}{431 \frac{\text{kg}}{\text{h}}} \times 100\%$$

$$= 99.9916\%$$

Because the DRE for the C_7H_8 is at least 99.99%, the C_7H_8 is in compliance.

The DRE for the C_8H_{10} is

$$DRE_{C_8} = \frac{\dot{m}_{in} - \dot{m}_{out}}{\dot{m}_{in}} \times 100\%$$

$$= \frac{434 \frac{kg}{h} - 0.07 \frac{kg}{h}}{434 \frac{kg}{h}} \times 100\%$$

$$= 99.9839\%$$

Because the DRE for the C_7H_8 is less than 99.99%, the C_8H_{10} is not in compliance.

The HCl emission may not exceed 1.8 kg/h or 1% of the HCl prior to the control equipment, whichever is greater. The emission of 1.1 kg/h in the table given in the problem statement meets the 1.8 kg/h limit. This is sufficient to demonstrate compliance. For completeness, the calculation of the mass emission rate prior to control follows.

The molecular weight of C_6H_5Cl, MW_{C_6}, is given as 112.5 g/mol. The molar feed rate of C_6H_5Cl is

$$F_{C_6} = \frac{\dot{m}_{in,C_6}}{MW_{C_6}} = \left(\frac{154 \frac{kg}{h}}{112.5 \frac{g}{mol}}\right)\left(1000 \frac{g}{kg}\right)$$

$$= 1369 \text{ mol/h}$$

Each molecule of C_6H_5Cl contains one atom of chlorine; therefore,

$$F_{HCl} = F_{C_6} = 1369 \text{ mol/h}$$

The molecular weight of HCl, MW_{HCl}, is given as 36.5 g/mol. The mass flow rate of HCl out of the incinerator stack is

$$\dot{m}_{HCl} = (MW_{HCl})F_{out,C_6} = \left(36.5 \frac{g}{mol}\right)\left(1369 \frac{mol}{h}\right)$$

$$= 49\,969 \text{ g/h} \quad (49.969 \text{ kg/h})$$

This is the HCl emission prior to control. 1% of the uncontrolled emission is

$$\dot{m}_U = 0.01 \dot{m}_{HCl} = (0.01)\left(49.969 \frac{kg}{h}\right)$$

$$= 0.49969 \text{ kg/h}$$

The emission of 1.1 kg/h is greater than 1% of the uncontrolled emission. However, the incinerator meets the HCl limit because the HCl emission is less than 1.8 kg/h.

The answer is (D).

Why Other Options Are Wrong

(A) This incorrect answer reports a contradictory solution to the problem. The solution to this problem demonstrates that C_7H_8 is in compliance.

(B) This incorrect answer reports a reverse solution to the problem. The solution to this problem demonstrates that C_6H_5Cl and HCl are in compliance.

(C) This incorrect answer reports a conflicting solution to the problem. The solution to this problem demonstrates that C_6H_5Cl and C_7H_8 are in compliance.

SOLUTION 91

The atmospheric pressure, p, is 14.7 lbf/in². The temperature of the nitrogen is

$$T = 90°F + 460°$$
$$= 550°R$$

The universal gas constant, R, is 10.73 lbf-ft³/lbmol-in²-°R. The molecular weight of nitrogen, MW, is 28 lbm/lbmol. The volume of the reactor, V_R, is given as 400 ft³. Assuming that the ambient air is 21% (by volume) oxygen, the initial concentration of oxygen in the reactor, C_0, is 0.21 lbmol/ft³. The final concentration of the oxygen in the reactor, C, is given as 0.04 lbmol/ft³. The oxygen mole balance is

$$\frac{dCV_R}{dt} = Q_iC_i - QC$$

For constant volume and constant density,

$$\frac{dCV_R}{dt} = Q_iC_i - QC$$

Integrating,

$$\frac{V_R}{Q}\int_{C_0}^{C} \frac{dC}{C_i - C} = \int_0^t dt$$

$$\ln \frac{C_i - C_0}{C_i - C} = \frac{Qt}{V_R}$$

Because there is no oxygen in the feed stream,

$$C_i = 0$$

$$\ln \frac{C_0}{C} = \frac{Qt}{V_R}$$

Solving for the total volume of pure nitrogen feed in the feed stream,

$$Qt = V_R \ln \frac{C_0}{C}$$

$$= (400 \text{ ft}^3) \ln \frac{0.21 \frac{lbmol}{ft^3}}{0.04 \frac{lbmol}{ft^3}}$$

$$= 663.3 \text{ ft}^3$$

Because the nitrogen is to be treated as an ideal gas, the mass of nitrogen is

$$m = \frac{Vp(\text{MW})}{RT}$$

$$= \frac{(663.3 \text{ ft}^3)\left(14.7 \dfrac{\text{lbf}}{\text{in}^2}\right)\left(28 \dfrac{\text{lbm}}{\text{lbmol}}\right)}{\left(10.73 \dfrac{\text{lbf-ft}^3}{\text{lbmol-in}^2\text{-}°\text{R}}\right)(550°\text{R})}$$

$$= 46.3 \text{ lbm} \quad (50 \text{ lbm})$$

The answer is (B).

Why Other Options Are Wrong

(A) This incorrect answer uses the volume of the reactor instead of the volume of nitrogen when calculating the mass of nitrogen.

(C) This incorrect answer uses the temperature in degrees Fahrenheit instead of using the absolute scale when calculating the mass of nitrogen.

(D) This answer uses the universal gas constant with incorrect units (0.73 atm-ft^3/lbmol-in^2-°R) when calculating the mass of nitrogen.

SOLUTION 92

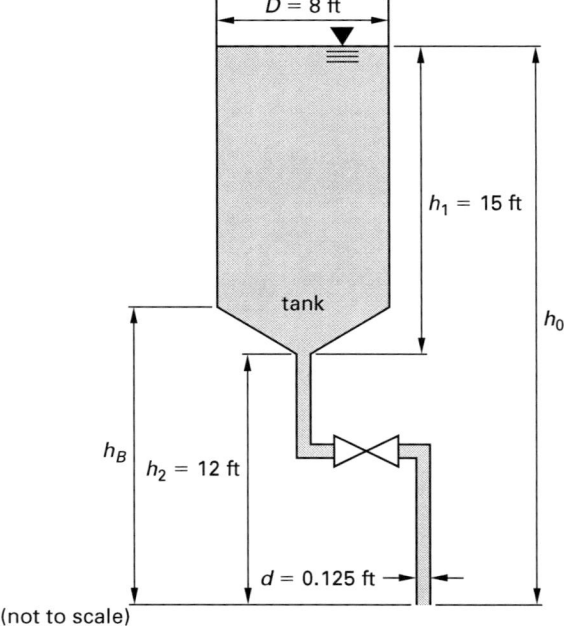

(not to scale)

The length of the pipe, L, is given as 200 ft. The elbow equivalent length, L_{elbow}, is given as 14 ft. The angle valve equivalent length, L_{valve}, is given as 55 ft. Because the pipe entrance shows a gradual change in the cross section and shape of the flow, the pipe entrance equivalent length applies.

$$K_{\text{entrance}} = 0.5$$

Because the pipe exit shows a sudden change in the cross section and shape of the flow, the pipe exit equivalent length applies.

$$K_{\text{exit}} = 1.00$$

The Darcy friction factor, f, is given as 0.03160. The equivalent length for the entire system is

$$K = f\left(\frac{L + 2L_{\text{elbows}} + L_{\text{valve}}}{d}\right) + K_{\text{entrance}} + K_{\text{exit}}$$

$$= (0.03160)\left(\frac{200 \text{ ft} + (2)(14 \text{ ft}) + 55 \text{ ft}}{0.125 \text{ ft}}\right)$$

$$+ 0.5 + 1.00$$

$$= 73.04$$

The height of the conical section, h_{con}, is given as 2.31 ft. The height of the bottom of the cylindrical portion of the tank above the drain-pipe outlet is

$$h_B = h_2 + h_{\text{con}} = 12 \text{ ft} + 2.31 \text{ ft}$$

$$= 14.31 \text{ ft}$$

The initial liquid height at time zero above the drain-pipe outlet is

$$h_o = h_2 + h_1 = 12 \text{ ft} + 15 \text{ ft}$$

$$= 27 \text{ ft}$$

The time required to drain the vertical cylindrical tank portion is given as

$$t_{\text{cyl,minutes}} = 24.058\sqrt{1+K}$$

$$= 24.058\sqrt{1+73.04}$$

$$= 207.01 \text{ min}$$

The time required to drain the vertical conical tank portion is given as

$$t_{\text{con,minutes}} = 1.768\sqrt{1+K}$$

$$= 1.768\sqrt{1+73.04}$$

$$= 15.21 \text{ min}$$

The total time required to drain the tank is

$$t_{\text{total}} = t_{\text{cyl}} + t_{\text{con}}$$

$$= 207.01 \text{ min} + 15.21 \text{ min}$$

$$= 222.2 \text{ min}$$

The volume of the cylindrical portion of the tank is

$$V_{\text{cyl}} = \left(\frac{\pi}{4}\right)D^2(h_o - h_B)$$

$$= \left(\frac{\pi}{4}\right)(8 \text{ ft})^2(27 \text{ ft} - 14.31 \text{ ft})$$

$$= 637.9 \text{ ft}^3$$

The volume of the conical portion of the tank, V_{con}, is given as 157.2 ft^3. The total volume to be drained is

$$V_{\text{total}} = V_{\text{cyl}} + V_{\text{con}} = 637.9 \text{ ft}^3 + 157.2 \text{ ft}^3$$
$$= 795.1 \text{ ft}^3$$

The pipe cross-sectional area is

$$A = \frac{\pi d^2}{4} = \frac{\pi (0.125 \text{ ft})^2}{4}$$
$$= 0.01227 \text{ ft}^2$$

The average fluid velocity is

$$\text{v} = \frac{V_{\text{total}}}{A t_{\text{total}}}$$
$$= \frac{795.1 \text{ ft}^3}{(0.01227 \text{ ft}^2)(222.2 \text{ min})\left(60\,\frac{\text{sec}}{\text{min}}\right)}$$
$$= 4.86 \text{ ft/sec} \quad (4.9 \text{ ft/sec})$$

The answer is (B).

Why Other Options Are Wrong

(A) This incorrect answer omits the volume of the conical portion of the tank when calculating the total volume.

(C) This incorrect answer excludes the time to drain the conical portion when calculating the total time.

(D) This incorrect answer uses the time in minutes instead of seconds when calculating the average velocity.

SOLUTION 93

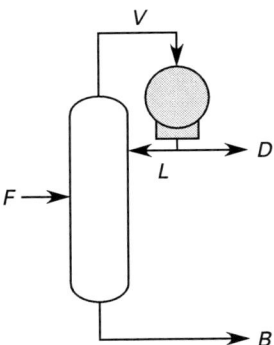

The mole fraction of trichloroethene in the feed, x_F, is given as 0.40. The mole fraction of trichloroethene in the distillate, x_D, is given as 0.96. The reflux ratio is given as

$$R = 2.1 = \frac{L}{D}$$

The rectifying operating line or R-line equation, assuming constant molar overflow, is

$$y = \left(\frac{R}{R+1}\right)x + \left(\frac{1}{R+1}\right)x_D$$
$$= \left(\frac{2.1}{2.1+1}\right)x + \left(\frac{1}{2.1+1}\right)0.96$$
$$= 0.677x + 0.3097$$

At a mole fraction of 0.40 in the liquid phase, the mole fraction in the vapor phase on the operating line is

$$y = (0.677)(0.40) + 0.3097$$
$$= 0.5805 \quad (0.58)$$

Because the feed is a saturated liquid, the q-line is a vertical line. The q-line represents the fraction of liquid in the feed. The q-line, x_F, is at 0.40.

The relative volatility, α, is given as 2.65. The equilibrium curve is

$$y = \frac{\alpha x}{1 + (\alpha - 1)x}$$
$$x = \frac{y}{\alpha + (1-\alpha)y} = \frac{y}{2.65 + (1 - 2.65)y}$$
$$= \frac{y}{2.65 + (-1.65)y}$$

In the following table, this equation is used to calculate the values in the x column.

The rectifying operating line, $y = 0.677x + 0.3097$, is used in the following table to calculate the values of the mole fraction in the vapor phase of trichloroethene. These values are in the y column of the following table before the feed plate.

plate	$x = \dfrac{y}{2.65 - 1.654}$	$y = 0.677x + 0.3097$
1	0.901	0.96
1	0.901	0.92
2	0.813	0.92
2	0.813	0.86
3	0.699	0.86
3	0.699	0.783
4	0.578	0.783
4	0.578	0.701
5	0.47	0.701
5	0.47	0.628
6	0.39	0.628 feed plate
6	0.39	0.574

Counting from the top of the colum, the feed plate that will achieve this separation is 6.

The graphical solution follows.

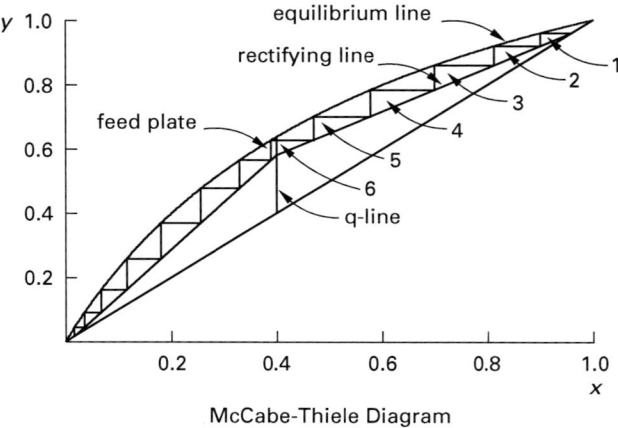

McCabe-Thiele Diagram

The answer is (B).

Why Other Options Are Wrong

(A) In this incorrect answer, the feed enters the column one plate higher than is optimum for the separation. The liquid fraction of trichloroethene at the feed plate would be 0.47 instead of the required 0.40.

(C) In this incorrect answer, the feed enters the column one plate lower than is optimum for the separation. The liquid fraction of trichloroethene would be 33% instead of the required 40%.

(D) In this incorrect answer, the feed enters the column two plates lower than is optimum for the separation. The liquid fraction of trichloroethene would be 26% instead of 40%.

SOLUTION 94

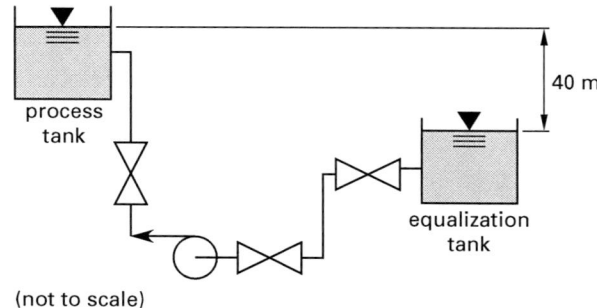

(not to scale)

The flow rate, Q, is given as $0.05 \text{ m}^3/\text{s}$. The normal service velocity, v_n, is given as 2.1 m/s. The internal area of the pipe is

$$A_n = \frac{\pi D^2}{4} = \frac{Q}{v_n}$$

Calculate the pipe size or the diameter of the pipe by dividing the volumetric flow rate by the desired velocity. The diameter of the pipe is

$$D = \sqrt{\frac{4Q}{\pi v_n}} = \sqrt{\frac{(4)\left(0.05 \dfrac{\text{m}^3}{\text{s}}\right)}{\pi \left(2.1 \dfrac{\text{m}}{\text{s}}\right)}}$$

$$= 0.174 \text{ m} \quad (174 \text{ mm})$$

From a standard pipe dimensions table, select 150 mm as the actual pipe size.

$$D_i = 0.150 \text{ m}$$

This value is the closest lower pipe diameter in the table. With this diameter, calculate the actual velocity in the pipe.

$$v = \frac{Q}{A} = \frac{Q}{\dfrac{\pi D_i^2}{4}} = \frac{0.05 \dfrac{\text{m}^3}{\text{s}}}{\dfrac{\pi (0.150 \text{ m})^2}{4}}$$

$$= 2.83 \text{ m/s}$$

The kinematic viscosity, ν, is given as $9.01 \times 10^{-7} \text{ m}^2/\text{s}$. The Reynolds number is

$$\text{Re} = \frac{D_i v}{\nu} = \frac{(0.150 \text{ m})\left(2.83 \dfrac{\text{m}}{\text{s}}\right)}{9.01 \times 10^{-7} \dfrac{\text{m}^2}{\text{s}}}$$

$$= 4.71 \times 10^5$$

The roughness, ϵ, is given as 1.4×10^{-6} m. The friction factor using the Colebrook and White equation is

$$\frac{1}{\sqrt{f}} = -2\log_{10}\left(\frac{\epsilon}{3.7 D_i} + \frac{2.51}{\text{Re}\sqrt{f}}\right)$$

$$= -2\log\left(\frac{1.4 \times 10^{-6} \text{ m}}{(3.7)(0.15 \text{ m})} + \frac{2.51}{4.71 \times 10^5 \sqrt{f}}\right)$$

The friction factor, f, is 0.01343. The sum of the resistance coefficient is

minor loss	resistance coefficient	number of components	resistance coefficient times number of components
entry	0.5	1	0.5
gate valve	0.2	1	0.2
swing check valve	2.5	2	5.0
elbow	0.3	4	1.4
exit	1.0	1	1.0
		total	8.1

$$\Sigma K = 8.1$$

The acceleration due to gravity, g, is 9.81 m/s².

Using the Darcy-Weisbach equation, calculate the pressure drop through the piping. The pipe length, L, is given as 70 m. The friction factor using the Darcy-Weisbach equation is

$$h_L = \left(\frac{fL}{D_i} + \Sigma K\right)\left(\frac{v^2}{2g}\right)$$

$$= \left(\frac{(0.01343)(70 \text{ m})}{0.150 \text{ m}} + 8.1\right)\left(\frac{\left(2.83 \frac{\text{m}}{\text{s}}\right)^2}{(2)\left(9.81 \frac{\text{m}}{\text{s}^2}\right)}\right)$$

$$= 5.86 \text{ m}$$

The static head, h_s, is given as 40 m. The pump head is the sum of the elevation head and the pressure drop through the pipe.

$$h_{\text{pump}} = h_s + h_L = 40 \text{ m} + 5.86 \text{ m}$$
$$= 45.86 \text{ m} \quad (46 \text{ m})$$

The answer is (C).

Why Other Options Are Wrong

(A) This incorrect answer uses the normal service velocity instead of the actual velocity calculated with the actual pipe diameter.

(B) This incorrect answer fails to use the resistance coefficients for both the exit and entrance when calculating the sum of the total resistance coefficient.

(D) This incorrect answer fails to use the gravity conversion factor when calculating the head loss due to friction.

SOLUTION 95

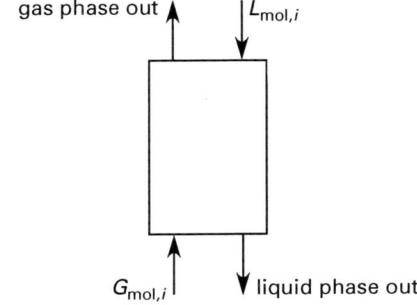

The Eckert modification to the generalized correlation for randomly packed towers based on flooding considerations is used to obtain the superficial gas flow rate entering the absorber, $G_{\text{sfr},i}$, and the $L_{\text{mol},i}/G_{\text{mol},i}$ ratio at the tower flood point.

The abscissa, from the plot of generalized pressure drop correlation, is

$$X = \left(\frac{L_{\text{mol},i}}{G_{\text{mol},i}}\right)\left(\frac{\text{MW}_L}{\text{MW}_G}\right)\sqrt{\frac{\rho_G}{\rho_L - \rho_G}}$$

$$= \frac{\left(7275 \frac{\text{lbmol}}{\text{hr}}\right)\left(18 \frac{\text{lbm}}{\text{lbmol}}\right)}{\left(3194 \frac{\text{lbmol}}{\text{hr}}\right)\left(29 \frac{\text{lbm}}{\text{lbmol}}\right)}$$

$$\times \sqrt{\frac{0.0709 \frac{\text{lbm}}{\text{ft}^3}}{62.4 \frac{\text{lbm}}{\text{ft}^3} - 0.0709 \frac{\text{lbm}}{\text{ft}^3}}}$$

$$= 0.0476815$$

The ordinate, from the plot of generalized pressure drop correlation, is given as

$$Y = e^{\left(\begin{array}{c}-4.026-0.9895\ln X-0.0829(\ln X)^2\\+0.0324(\ln X)^3+0.0053(\ln X)^4\end{array}\right)}$$

$$= e^{\left(\begin{array}{c}-4.026-0.9895\ln 0.0476815-0.0829(\ln 0.0476815)^2\\+0.0324(\ln 0.0476815)^3+0.0053(\ln 0.0476815)^4\end{array}\right)}$$

$$= 0.106345 \text{ lbm-ft-cP}^{0.1}/\text{lbf}$$

The packing factor, F_p, is given as 65. The gravitational constant, g_c, is 32.17 ft-lbf/lbm-sec². The ordinate given in the problem statement is

$$Y = \frac{G_{\text{sfr},i}^2 F_p^{0.1} \mu_L}{(\rho_L - \rho_G)\rho_G g_c}$$

$$= 0.106345 \text{ lbm-ft-cP}^{0.1}/\text{lbf}$$

Solving for $G_{\text{sfr},i}$,

$$G_{\text{sfr},i} = \sqrt{\frac{(\rho_L - \rho_G)\rho_G g_c Y}{F_p \mu_L^{0.1}}}$$

$$= \sqrt{\frac{\begin{array}{c}\left(62.4 \frac{\text{lbm}}{\text{ft}^3} - 0.0709 \frac{\text{lbm}}{\text{ft}^3}\right)\left(0.0709 \frac{\text{lbm}}{\text{ft}^3}\right)\\ \times \left(32.17 \frac{\text{ft-lbf}}{\text{lbm-sec}^2}\right)\\ \times \left(0.106345 \frac{\text{lbm-ft-cP}^{0.1}}{\text{lbf}}\right)\end{array}}{(65)\left(\frac{2.16 \frac{\text{lbm}}{\text{ft-hr}}}{2.42 \frac{\text{lbm}}{\text{ft-hr-cP}}}\right)^{0.1}}}$$

$$= 0.485025 \text{ lbm/ft}^2\text{-sec}$$

The cross-sectional area of the column is

$$A = \frac{G_{\text{mol},i} \text{MW}_G}{G_{\text{sfr},i}}$$

$$= \frac{\left(3194 \, \frac{\text{lbmol}}{\text{hr}}\right)\left(29 \, \frac{\text{lbm}}{\text{lbmol}}\right)}{\left(3600 \, \frac{\text{sec}}{\text{hr}}\right)\left(0.485025 \, \frac{\text{lbm}}{\text{ft}^2\text{-sec}}\right)}$$

$$= 53.0477 \, \text{ft}^2$$

The diameter of the column is

$$D = \sqrt{\frac{4A}{\pi}} = \sqrt{\frac{(4)(53.0477 \, \text{ft}^2)}{\pi}}$$

$$= 8.21842 \, \text{ft} \quad (8.2 \, \text{ft})$$

The answer is (B).

Why Other Options Are Wrong

(A) This incorrect answer fails to use the exponent on the denominator when calculating the superficial gas flow rate.

(C) This incorrect answer fails to use the gravitational constant when calculating the superficial gas flow rate.

(D) This incorrect answer fails to use the conversion factor from seconds to hours when calculating the cross-sectional area.

SOLUTION 96

The distillate flow rate is

$$D = D_{\text{ac}} + D_W = 1218 \, \frac{\text{mol}}{\text{min}} + 36 \, \frac{\text{mol}}{\text{min}}$$

$$= 1254 \, \text{mol/min}$$

The reflux ratio, R_R, is given as 1.1. The superficial molar liquid velocity is

$$L = R_R D = (1.1)\left(1254 \, \frac{\text{mol}}{\text{min}}\right)$$

$$= 1379.4 \, \text{mol/min}$$

A mole balance around the splitter gives the velocity of the gas through the net tower area. The velocity of the gas is

$$V = L + D = 1379.4 \, \frac{\text{mol}}{\text{min}} + 1254 \, \frac{\text{mol}}{\text{min}}$$

$$= 2633.4 \, \text{mol/min}$$

The pressure drop through the column, Δp, is 50 kPa. The pressure at the bottom of the column is

$$p_2 = p_1 + \Delta p = 105 \, \text{kPa} + 50 \, \text{kPa}$$

$$= 155 \, \text{kPa}$$

The universal gas constant, R, is 8.314 J/mol·K. The density of the liquid phase, ρ_L, is given as 10^6 g/m^3. The molecular weight of the vapor phase, MW, is given as 18 g/mol. Because the vapor is an ideal gas, the density of the vapor for the bottom of the column is

$$\rho_g = \frac{p_2(\text{MW})}{RT} = \frac{(155 \, \text{kPa})\left(1000 \, \frac{\text{Pa}}{\text{kPa}}\right)\left(18 \, \frac{\text{g}}{\text{mol}}\right)}{\left(8.314 \, \frac{\text{J}}{\text{mol·K}}\right)(382 \, \text{K})}$$

$$= 878.5 \, \text{g/m}^3$$

A mole balance around the distillation column gives the molar feed rate. The molar feed rate is

$$F = D_{\text{ac}} + D_W + B_{\text{ac}} + B_W$$

$$= 1218 \, \frac{\text{mol}}{\text{min}} + 36 \, \frac{\text{mol}}{\text{min}} + 282 \, \frac{\text{mol}}{\text{min}} + 20\,064 \, \frac{\text{mol}}{\text{min}}$$

$$= 21\,600 \, \text{mol/min}$$

Because the feed is at its bubble point, the mass flow rate of the liquid in the bottom section is

$$\dot{m}_{L'} = (L + F)(\text{MW})$$

$$= \left(1379.4 \, \frac{\text{mol}}{\text{min}} + 21\,600 \, \frac{\text{mol}}{\text{min}}\right)\left(18 \, \frac{\text{g}}{\text{mol}}\right)$$

$$= 413\,629 \, \text{g/min}$$

The mass flow rate of the vapor in the bottom section is

$$\dot{m}_{V'} = V(\text{MW}) = \left(2633.4 \, \frac{\text{mol}}{\text{min}}\right)\left(18 \, \frac{\text{g}}{\text{mol}}\right)$$

$$= 47\,401 \, \text{g/min}$$

The surface tension, σ, is given as 70 dyn/cm. The abscissa of the flooding curve is given as

$$F_{\text{lv}} = \frac{\dot{m}_L}{\dot{m}_{V'}}\sqrt{\frac{\rho_g}{\rho_L}} = \left(\frac{413\,629 \, \frac{\text{g}}{\text{min}}}{47\,401 \, \frac{\text{g}}{\text{min}}}\right)\sqrt{\frac{878.5 \, \frac{\text{g}}{\text{m}^3}}{10^6 \, \frac{\text{g}}{\text{m}^3}}}$$

$$= 0.259$$

For a 24 in tray spacing, the ordinate from the flooding curve is given as

$$C_{\text{sb}} = e^{-3.46 + \sqrt{1.69 - 2.03 \ln F_{\text{lv}} - 0.217(\ln F_{\text{lv}})^2}}$$

$$= e^{-3.46 + \sqrt{1.69 - 2.03 \ln 0.259 - 0.217(\ln 0.259)^2}}$$

$$= 0.234 \, \text{m/s}$$

The ordinate for a 24 in tray spacing is given as

$$C_{\text{sb}} = U_{\text{nf}} \left(\frac{20 \times 10^{-2} \frac{\text{N}}{\text{m}}}{\sigma} \right)^{0.2} \sqrt{\frac{\rho_g}{\rho_L - \rho_g}}$$

Solving for the flooding velocity,

$$U_{\text{nf}} = C_{\text{sb}} \left(\frac{\sigma}{20 \times 10^{-2} \frac{\text{N}}{\text{m}}} \right)^{0.2} \sqrt{\frac{\rho_L}{\rho_g} - 1}$$

$$= \left(0.234 \; \frac{\text{m}}{\text{s}}\right) \left(\frac{70 \times 10^{-2} \frac{\text{N}}{\text{m}}}{20 \times 10^{-2} \frac{\text{N}}{\text{m}}} \right)^{0.2} \sqrt{\frac{10^6 \; \frac{\text{g}}{\text{m}^3}}{878.5 \; \frac{\text{g}}{\text{m}^3}} - 1}$$

$$= 10.14 \; \text{m/s}$$

At 80% flooding, the velocity is

$$U = 0.80 U_{\text{nf}} = (0.80)\left(10.14 \; \frac{\text{m}}{\text{s}}\right)$$
$$= \left(8.11 \; \frac{\text{m}}{\text{s}}\right)\left(60 \; \frac{\text{s}}{\text{min}}\right)$$
$$= 486.6 \; \text{m/min}$$

The fraction of the area available for vapor flow, ϵ, is given as 0.60. The diameter of the section below the feed tray is

$$d = 2 \sqrt{\frac{\dot{m}_{V'}}{\rho_g U \epsilon \pi}}$$

$$= (2) \sqrt{\frac{47\,401 \; \frac{\text{g}}{\text{min}}}{\left(878.5 \; \frac{\text{g}}{\text{m}^3}\right)\left(486.6 \; \frac{\text{m}}{\text{min}}\right)(0.6)\pi}}$$

$$= 0.485 \; \text{m} \quad (0.49 \; \text{m})$$

The answer is (C).

Why Other Options Are Wrong

(A) This incorrect answer does not take the square root when calculating the diameter.

(B) This incorrect answer does not use the fraction of the area available for vapor flow when calculating the diameter.

(D) This incorrect answer uses the flooding velocity in meters per second instead of meters per minute when calculating the diameter.

SOLUTION 97

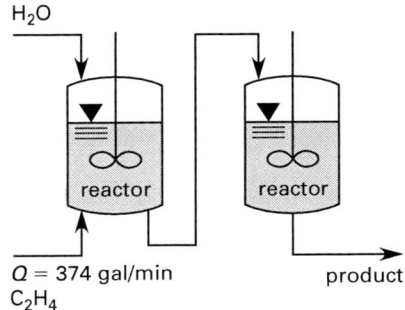

The temperature of the reaction is given as

$$T = 30°\text{C} + 273° = 303\text{K}$$

The conversion of C_2H_4 is

$$X = (0.0012\text{K}^{-1})T + 0.5$$
$$= (0.0012\text{K}^{-1})(303\text{K}) + 0.5$$
$$= 0.864$$

The reaction constant is

$$k = \left(-0.0145 \; \frac{1}{\text{K-min}}\right) T + 9.682 \; \text{min}^{-1}$$
$$= \left(-0.0145 \; \frac{1}{\text{K-min}}\right)(303\text{K}) + \frac{9.682}{\text{min}}$$
$$= 5.29 \; \text{min}^{-1}$$

For two CSTRs in series with equal volumes and a first-order reaction, the material balance is

$$(k\tau + 1)^2 = \frac{1}{1-X}$$

Solving for the space time, τ, gives

$$\tau = \frac{\sqrt{\frac{1}{1-X}} - 1}{k} = \frac{\sqrt{\frac{1}{1-0.864}} - 1}{5.29 \; \text{min}^{-1}}$$
$$= 0.3236 \; \text{min}$$

The volume of the reactor is

$$V = Q\tau = \left(374 \; \frac{\text{gal}}{\text{min}}\right)(0.3236 \; \text{min})$$
$$= 121.026 \; \text{gal} \quad (120 \; \text{gal})$$

The answer is (B).

Why Other Options Are Wrong

(A) This incorrect answer uses the temperature in degrees Celsius instead of the absolute scale when calculating the reaction rate constant and the conversion.

(C) This incorrect answer omits the exponent 2 when calculating the space time.

(D) This incorrect answer divides by the space time instead of multiplying when calculating the volume of the reactor.

SOLUTION 98

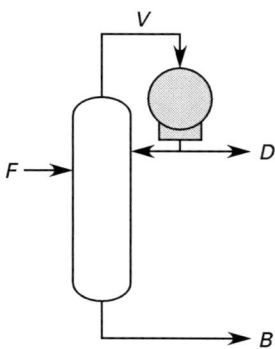

The basis of calculation is the feed molar flow rate of 1 lbmol/hr. The feed molar flow rate, F, is 1 lbmol/hr. The mole fraction of C1 in the feed, x_F, is given as 0.45. The mole fraction of C1 in the distillate, x_D, is given as 0.97. The mole fraction of C1 in the bottoms, x_B, is given as 0.03. The reflux ratio, R, is given as 2.81. A total mass balance around the distillation column gives

$$F = D + B$$

A mass balance of C1 around the distillation column gives

$$x_F F = x_D D + x_B B$$
$$0.45 F = 0.97 D + 0.03 B$$

Solving simultaneously gives

$$D = \left(\frac{x_F - x_B}{x_D - x_B}\right) F$$
$$= \left(\frac{0.45 - 0.03}{0.97 - 0.03}\right) \left(1 \; \frac{\text{lbmol}}{\text{hr}}\right)$$
$$= 0.447 \; \text{lbmol/hr}$$

$$B = F - D = 1 \; \frac{\text{lbmol}}{\text{hr}} - 0.447 \; \frac{\text{lbmol}}{\text{hr}}$$
$$= 0.553 \; \text{lbmol/hr}$$

The rectifying operating line, or R-line, equation is

$$y = \left(\frac{R}{R+1}\right) x + \left(\frac{1}{R+1}\right) x_D$$
$$= \left(\frac{2.81}{2.81+1}\right) x + \left(\frac{1}{2.81+1}\right) (0.97)$$
$$= 0.738x + 0.255$$

At a mole fraction of 0.45 in the liquid phase, the mole fraction in the vapor phase is

$$y = (0.738)(0.45) + 0.255$$
$$= 0.5871 \quad (0.59)$$

Because the feed is a saturated liquid, the q-line is a vertical line. The q-line represents the fraction of liquid in the feed. The q-line is at

$$x_F = 0.45$$

Below the feed plate, the following relationships hold

$$R \equiv \frac{L}{D}$$

Therefore,

$$L_S = F + DR = 1 \; \frac{\text{lbmol}}{\text{hr}} + \left(0.447 \; \frac{\text{lbmol}}{\text{hr}}\right)(2.81)$$
$$= 2.256 \; \text{lbmol/hr}$$

The stripping operating line, or S-line, equation is

$$y = \left(\frac{L_S}{L_S - B}\right) x - \frac{B x_B}{L_S - B}$$
$$= \left(\frac{2.256}{2.256 - 0.553}\right) x - \frac{(0.553)(0.03)}{2.256 - 0.553}$$
$$= 1.325 x - 0.0097$$

The coordinates of end points of operating lines are given in the following table.

line	x	y	comes from
R-line	0.97	0.97	$y = x$ line
R-line	0.45	0.59	x from the rectification line equation
S-line	0.45	0.59	y from the stripping line equation
S-line	0.03	0.03	$y = x$ line

The relative volatility, α, is given as 3.2. The equilibrium curve is

$$x = \frac{y}{\alpha + (1 - \alpha) y} = \frac{y}{3.2 + (1 - 3.2) y}$$
$$= \frac{y}{3.2 + (-2.2) y}$$

In the following table, this equation is used to calculate the values in the x column.

The rectifying operating line, $y = 0.738x + 0.255$, is used in the following table to calculate the values of the mole fraction in the vapor phase of C1. These values are in the y column of the following table, before the feed plate.

The following equation is used to calculate the values in the y column of the following table, after the feed plate.

$$y = 1.325x - 0.0097$$

plate	x	y	
1	0.97	0.97	
1	0.91	0.97	
2	0.91	0.927	
2	0.796	0.927	
3	0.796	0.842	
3	0.624	0.842	
4	0.624	0.716	feed plate
4	0.439	0.716	
5	0.439	0.572	
5	0.295	0.572	
6	0.295	0.381	
6	0.161	0.381	
7	0.161	0.204	
7	0.074	0.204	
8	0.074	0.088	bottom plate
8	0.029	0.088	

The number of plates needed is 8.

The graphical solution follows.

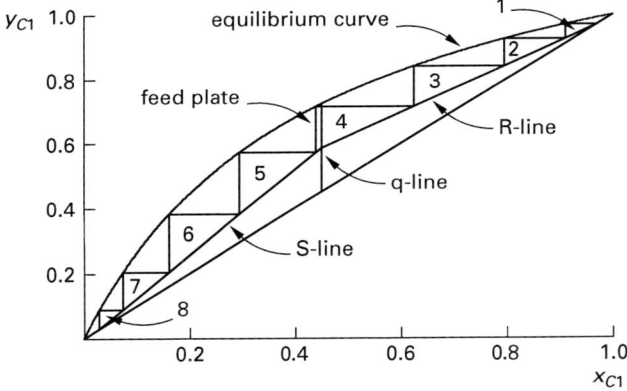

The answer is (C).

Why Other Options Are Wrong

(A) This incorrect answer reports the feed plate as the plate where the separation is achieved.

(B) This incorrect answer reports one plate less than is necessary to achieve the separation. The bottoms would have a concentration of 0.074 larger than the concentration required of 0.03 for the C2 in the residue produced.

(D) This incorrect answer reports one plate more than is required to achieve the separation requested. The bottoms would contain less than 3% C1.

SOLUTION 99

The concentration of COD in stream 1, C_{COD1}, is given as 0.09 kg/m³. The concentration of soluble COD in stream 1, $C_{S,COD1}$, is given as 0.11 kg/m³. Because the clarifier overflow contains no solids, the concentration of COD in stream 3, C_{COD3}, is 0 kg/m³.

The clarifier overflow soluble suspended solids concentration, $C_{S,COD3}$, is given as 0.032 kg/m³. The clarifier underflow concentration of COD in stream 4, C_{COD4}, is given as 9 kg/m³. The fraction of biological solids converted to biomass, calculated on a COD basis, Y, is given as 0.53. The specific biomass loss rate for solids formed in the basin, b_1, is given as 0.034 d⁻¹. The specific biomass loss rate for biomass fed, b_2, is given as 0.034 d⁻¹.

The mean biological solids retention time, θ, is given as 2.8 d. The mass of biological solids in the aeration basin is VC_{COD2}. The mass per unit time of solids removed as intended wastage is $Q_4 C_{COD4}$. The mass per unit time of solids removed in the clarifier overflow is $Q_3 C_{COD3}$. Assuming no significant evaporation, the mass balance around the wastewater treatment gives

$$Q_3 = Q_1 - Q_4$$

Replacing in the preceding equation gives

$$Q_3 C_{COD3} = (Q_1 - Q_4) C_{COD3}$$

A mass balance for the mass of microorganisms in the basin is

rate of accumulation of microorganism within the system boundary	=	rate of flow of microorganism into the system boundary

	rate of flow of microorganism out of the system boundary	+	net growth of microorganism within the boundary
−			

accumulation = inflow − outflow + net growth

$$\frac{dC_{COD}}{dt} V = Q_1 C_{COD1} \\ - ((Q_1 - Q_4) C_{COD3} - Q_4 C_{COD4}) \\ + r_g V$$

Assuming that the concentration of microorganisms in the influent is negligible,

$$C_{COD1} = 0 \text{ kg/m}^3$$

Assuming that steady-state conditions prevail,

$$\frac{dC_{COD}}{dt} = 0$$

Replacing,
$$(Q_1 - Q_4)C_{COD3} + Q_4 C_{COD4} = r_g V$$

Divide both sides of the equation by VC_{COD4}.
$$\frac{(Q_1 - Q_4)C_{COD3} + Q_4 C_{COD4}}{VC_{COD2}} = \frac{r_g V}{VC_{COD4}}$$

The inverse of the term on the left-hand side is defined as the mean biological solids retention time.
$$\theta = \frac{VC_{COD2}}{(Q_1 - Q_4)C_{COCD3} + Q_4 C_{COD4}} \quad [I]$$

The mass of solids per unit time exiting the reactor due to biomass generated is $\dot{m}_1$. By definition, the specific biomass loss rate for solids formed in the basin is
$$b_1 = \frac{Q_1 Y(C_{S,COD1} - C_{S,COD3}) - \dot{m}_1}{\dot{m}_1 \theta}$$

Solving for the mass of solids per unit time exiting the reactor due to biomass generated,
$$\dot{m}_1 = \frac{Q_1 Y(C_{S,COD1} - C_{S,COD3})}{1 + b_1 \theta} \quad [II]$$

The mass of solids per unit time exiting the reactor due to solids fed is $\dot{m}_2$. By definition, the specific biomass loss rate for biomass in the feed is
$$b_2 = \frac{Q_1 C_{COD1} - \dot{m}_2}{\dot{m}_2 \theta}$$

Solving for the mass of solids per unit time exiting the reactor due to solids fed,
$$\dot{m}_2 = \frac{Q_1 C_{COD1}}{1 + b_2 \theta} \quad [III]$$

Solving for θ and replacing Eqs. II and III gives
$$\theta = \frac{VC_{COD2}}{\dot{m}_1 + \dot{m}_2}$$
$$= \frac{VC_{COD2}}{\frac{Q_1 Y(C_{S,COD1} - C_{S,COD3})}{1 + b_1 \theta} + \frac{Q_1 C_{COD1}}{1 + b_2 \theta}}$$

Solving for Eq. I for the concentration of COD in stream 2 gives
$$C_{COD2} = \left(\frac{\theta Q_1}{V}\right)\left(\frac{Y(C_{S,COD1} - C_{S,COD3})}{1 + b_1 \theta} + \frac{C_{COD1}}{1 + b_2 \theta}\right)$$
$$= \left(\frac{(2.8 \text{ d})\left(132 \frac{\text{m}^3}{\text{d}}\right)}{12.1 \text{ m}^3}\right)$$
$$\times \left(\frac{(0.53)\left(0.11 \frac{\text{kg}}{\text{m}^3} - 0.032 \frac{\text{kg}}{\text{m}^3}\right)}{1 + (0.034 \text{ d}^{-1})(2.8 \text{ d})} + \frac{0.09 \frac{\text{kg}}{\text{m}^3}}{1 + (0.034 \text{ d}^{-1})(2.8 \text{ d})}\right)$$
$$= 3.663 \text{ kg/m}^3$$

Because there is no change in the clarifier overflow soluble suspended solids concentration as the water stream goes from the aeration basin to the clarifier, the overflow soluble suspended solids concentration at the exit of the aeration basin equals the clarifier soluble suspended solids concentration.

Solving Eq. I for the mass flow rate of stream 4 gives
$$Q_4 = \frac{\frac{VC_{COD2}}{\theta} - Q_1 C_{COD3}}{C_{COD4} - C_{COD3}}$$
$$= \frac{\frac{(12.1 \text{ m}^3)\left(3.663 \frac{\text{kg}}{\text{m}^3}\right)}{2.8 \text{ d}} - \left(132 \frac{\text{m}^3}{\text{d}}\right)\left(0 \frac{\text{kg}}{\text{m}^3}\right)}{9 \frac{\text{kg}}{\text{m}^3} - 0 \frac{\text{kg}}{\text{m}^3}}$$
$$= 1.759 \text{ m}^3/\text{d}$$

The mass balance around the wastewater treatment gives
$$Q_3 = Q_1 - Q_4$$

Replacing,
$$Q_3 = 132 \frac{\text{m}^3}{\text{d}} - 1.759 \frac{\text{m}^3}{\text{d}} = 130.24 \text{ m}^3/\text{d}$$

The mass balance of solids in the basin gives the concentration of biological solids produced in the aeration basin.
$$C_{COD} = Q_4 C_{COD4} + Q_3 C_{COD3} - \frac{Q_1 C_{COD1}}{1 + b_2 \theta}$$
$$= \left(1.759 \frac{\text{m}^3}{\text{d}}\right)\left(9 \frac{\text{kg}}{\text{m}^3}\right)$$
$$+ \left(130.24 \frac{\text{m}^3}{\text{d}}\right)\left(0 \frac{\text{kg}}{\text{m}^3}\right)$$
$$- \frac{\left(132 \frac{\text{m}^3}{\text{d}}\right)\left(0.09 \frac{\text{kg}}{\text{m}^3}\right)}{1 + (0.034 \text{ d}^{-1})(2.8 \text{ d})}$$
$$= 4.984 \text{ kg/d}$$

The oxygen mass balance gives the concentration of oxygen required in the process.
$$C_{O_2} = Q_1(C_{S,COD1} - C_{S,COD3})$$
$$- C_{COD} + Q_1 C_{COD1}\left(1 - \frac{1}{1 + b_2 \theta}\right)$$
$$= \left(132 \frac{\text{m}^3}{\text{d}}\right)\left(0.11 \frac{\text{kg}}{\text{m}^3} - 0.032 \frac{\text{kg}}{\text{m}^3}\right)$$
$$- 4.984 \frac{\text{kg}}{\text{m}^3} + \left(132 \frac{\text{m}^3}{\text{d}}\right)\left(0.09 \frac{\text{kg}}{\text{m}^3}\right)$$
$$\times \left(1 - \frac{1}{1 + (0.034 \text{ d}^{-1})(2.8 \text{ d})}\right)$$
$$= 6.345 \text{ kg/day} \quad (6.3 \text{ kg/day})$$

The answer is (B).

Why Other Options Are Wrong

(A) This incorrect answer subtracts the term representing the organic matter content of the feed instead of adding the term when calculating the oxygen required.

(C) This incorrect answer uses the opposite sign of the term representing the oxygen needed to oxidize the organic matter to produce the COD formed during the wastewater treatment.

(D) This incorrect answer adds when it should subtract the last term in the expression when calculating the concentration of oxygen required in the process. This answer uses

$$C_{O_2} = Q_1(C_{S,\text{COD}1} - C_{S,\text{COD}3}) - C_{\text{COD}} + Q_1 C_{\text{COD}1}$$
$$\times \left(1 + \frac{1}{1 + b_2\theta}\right)$$

SOLUTION 100

The molecular weights, the number of hydrogen ions transferred, and the equivalent weights of the species are

species	molecular weight, MW (g/mol)	number of hydrogen ions transferred, n	equivalent weight, EW = MW/n (g/mol)
$CaCO_3$	100	2	50
CO_3^{2-}	60	2	30
HCO_3^-	61	1	61
H^+	1	1	1
OH^-	17	1	17

Because the pH of the wastewater is given as 9, the concentration of H^+ is

$$[H^+] = 10^{-\text{pH}} = 10^{-9} \text{ mol/L}$$

$$H^+ = \left(10^{-9} \frac{\text{mol}}{\text{L}}\right)(\text{MW}_{H^+})$$

$$= \left(10^{-9} \frac{\text{mol}}{\text{L}}\right)\left(1 \frac{\text{g}}{\text{mol}}\right)\left(1000 \frac{\text{mg}}{\text{g}}\right)$$

$$= 10^{-6} \text{ mg/L}$$

At 25°C, the ion product of water, K_w, is given as 10^{-14} mol²/L².

The concentration of OH^- is

$$[OH^-] = \frac{K_w}{[H^+]} = \frac{10^{-14} \frac{\text{mol}^2}{\text{L}^2}}{10^{-9} \frac{\text{mol}}{\text{L}}}$$

$$= 10^{-5} \text{ mol/L}$$

$$C_{OH^-} = \left(10^{-5} \frac{\text{mol}}{\text{L}}\right)(\text{EW}_{OH^-})$$

$$= \left(10^{-5} \frac{\text{mol}}{\text{L}}\right)\left(17 \frac{\text{g}}{\text{mol}}\right)\left(1000 \frac{\text{mg}}{\text{g}}\right)$$

$$= 0.17 \text{ mg/L}$$

The concentration of CO_3^{2-} in the wastewater, $W_{CO_3^{2-}}$, is given as 125 mg/L. The concentration of CO_3^{2-} as $CaCO_3$ is

$$[CO_3^{2-}] = W_{CO_3^{2-}} \left(\frac{\text{EW}_{CaCO_3}}{\text{EW}_{CO_3^{2-}}}\right)$$

$$C_{CO_3^{2-}} = \left(125 \frac{\text{mg}}{\text{L}}\right)\left(\frac{50 \frac{\text{g}}{\text{mol}}}{30 \frac{\text{g}}{\text{mol}}}\right)$$

$$= 208.3 \text{ mg/L}$$

The concentration of HCO_3^- in the wastewater, $W_{HCO_3^-}$, is given as 82 mg/L. The concentration of HCO_3^- as $CaCO_3$ is

$$[HCO_3^-] = W_{HCO_3^-} \left(\frac{\text{EW}_{CaCO_3}}{\text{EW}_{HCO_3^-}}\right)$$

$$C_{HCO_3^-} = \left(82 \frac{\text{mg}}{\text{L}}\right)\left(\frac{50 \frac{\text{g}}{\text{mol}}}{61 \frac{\text{g}}{\text{mol}}}\right)$$

$$= 67.21 \text{ mg/L}$$

The concentration of H^+ as $CaCO_3$ is

$$[H^+] = H^+ \left(\frac{\text{EW}_{CaCO_3}}{\text{EW}_{H^+}}\right)$$

$$C_{H^+} = \left(10^{-6} \frac{\text{mg}}{\text{L}}\right)\left(\frac{50 \frac{\text{g}}{\text{mol}}}{1 \frac{\text{g}}{\text{mol}}}\right)$$

$$= 5 \times 10^{-5} \text{ mg/L}$$

The concentration of OH^- as $CaCO_3$ is

$$[OH^-] = OH^- \left(\frac{\text{EW}_{CaCO_3}}{\text{EW}_{OH^-}}\right)$$

$$C_{OH^-} = \left(0.17 \frac{\text{mg}}{\text{L}}\right)\left(\frac{50 \frac{\text{g}}{\text{mol}}}{17 \frac{\text{g}}{\text{mol}}}\right)$$

$$= 0.5 \text{ mg/L}$$

The carbonate alkalinity is defined as

$$\text{alkalinity} = C_{HCO_3^-} + 2C_{CO_3^{2-}} + C_{OH^-} - C_{H^+}$$

$$= 67.21 \frac{\text{mg}}{\text{L}} + (2)\left(208.3 \frac{\text{mg}}{\text{L}}\right)$$

$$+ 0.5 \frac{\text{mg}}{\text{L}} - 5 \times 10^{-5} \frac{\text{mg}}{\text{L}}$$

$$= 484.31 \text{ mg/L} \quad (484 \text{ mg/L})$$

The answer is (D).

Why Other Options Are Wrong

(A) This incorrect answer does not multiply the concentration of HCO_3^- by 2 when calculating the alkalinity.

(B) This incorrect answer multiplies the concentration of HCO_3^- by 2 instead of the concentration of CO_3^{2-} when calculating the alkalinity.

(C) This incorrect answer subtracts the concentration of HCO_3^- instead of adding it when calculating the alkalinity.

Turn to PPI for Your Chemical PE Exam Review Materials
The Most Trusted Source for Chemical PE Exam Preparation
Visit www.ppi2pass.com today!

The Most Comprehensive Reference Materials

Chemical Engineering Reference Manual for the PE Exam
Michael R. Lindeburg, PE

- The most widely used chemical PE exam reference
- 66 chapters provide in-depth review of exam topics
- More than 380 solved example problems
- Hundreds of key tables, charts, and graphs at your fingertips
- Full glossary for quick reference
- Quickly locate information through the complete index

Quick Reference for the Chemical Engineering PE Exam
Michael R. Lindeburg, PE

- Quickly and easily access the formulas needed most often during the exam
- Drawn from the *Chemical Engineering Reference Manual*
- Quickly retrieve formulas without the distraction of surrounding text
- Organized by topic and indexed for rapid retrieval

The Practice You Need to Succeed

Practice Problems for the Chemical Engineering PE Exam
Michael R. Lindeburg, PE

- The perfect companion to the *Chemical Engineering Reference Manual*
- 450 practice problems increase your problem-solving skills
- Multiple-choice format, just like the exam
- Coordinated with the *Chemical Engineering Reference Manual* for focused preparation
- Step-by-step solutions provide immediate feedback

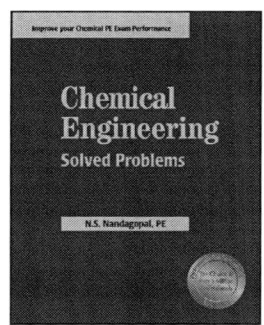

Chemical Engineering Solved Problems
N.S. Nandagopal, PE

- Collection of over 160 problems based on 26 scenarios
- Step-by-step solutions included for each problem
- Increase your speed and confidence
- Assess your strengths and weaknesses

Six-Minute Solutions for Chemical PE Exam Problems
Marta Vasquez, PhD, PE, and Robert R. Zinn

- Learn to solve problems in under 6 minutes
- 100 multiple-choice problems with solutions
- Improve your problem-solving speed and skills
- Perfect for the breadth and depth portions
- Discover how to avoid common mistakes

For the latest chemical PE exam news, the latest test-taker advice, the unique community of the Exam Forum, and FAQs, go to www.ppi2pass.com.

The Power to Pass™
www.ppi2pass.com